Potato developments in a changing Europe

Potato developments in a changing Europe

edited by:
Norbert U. Haase
Anton J. Haverkort

Wageningen Academic
Publishers

ISBN-10: 90-8686-011-7
ISBN-13: 978-90-8686-011-1

First published, 2006

Wageningen Academic Publishers
The Netherlands, 2006

Advisory committee Potato Europe
4-6 September 2006, Hameln, Germany

- Dr. B.V. Anisimov, Russian Potato Research Institute, Russia
- Dr. H. Böhm, Europlant Pflanzenzucht GmbH, Germany
- F. Bork, Grimme Landmaschinenfabrik GmbH & Co KG, Germany
- J. Bruer, Stöver Produktions GmbH & Co KG, Germany
- Dr. A. von Chappuis, DLG, Germany
- Dr. S.C. Chiru, Institute for Potato and Sugarbeet, Romania
- Dykes, British Potato Council, United Kingdom
- J.E. Godfrey, R.J. & A.E. Godfrey, United Kingdom
- Prof. Dr. J.F. Guenthner, Idaho Center for Potato Research & Education, Univ. of Idaho, USA
- J.H.J. Haarhuis, Agricultural Promotion Projects, the Netherlands
- Prof. Dr. A. Hanafi, Institut Agronomique et Veterinaire Hassan II, Morocco
- Dr. A.J. Haverkort, Plant Research International, the Netherlands
- Dr. M. Huarte, National Potato Program Coordinator, INTA Balcarce, Argentina
- Eng. G.G. Meiborg, Farm Frites, Poland
- Dr. R. Peters, KTBL-Versuchsstation Dethlingen, Germany
- L. Pieterse, Global Potato Focus Inc., Canada
- H.W. Platt, Agriculture & Agri-Food, Canada
- J. Procházka, Europlant šlechtitelská spol.s.r.o., Czech Republic
- Dr. E.R. Ritter, Neiker, Spain
- M. Umhau, Oschatz (Sachsen), Germany
- Prof. dr. G. Wenzel, Technische Universität München, Germany
- Prof. Dr. E. Zimnoch-Guzowska, The Plant Breeding and Acclimatization Institute, Mlochów Research Center, Poland

Organisers

- DLG - Deutsche Landwirtschafts-Gesellschaft e.V.
- Europoint b.v.

Table of contents

Crop protection and fertilisation

Product quality and certification

Trade of seed and ware potatoes

Consumer behaviour and marketing

Preface

The Potato 2006 congress was held on September 3-6 2006 in Hameln near Hannover and dealt largely with developments in the potato industry in an ever changing Europe. The European Union was just expanded recently with ten additional countries of which many are major potato producing countries such as the Baltic States and Poland. These countries underwent major changes in the potato industry due to a reduced demand for potato as feed and due to altering consumer requirements with major European retail chains which entered the market. Other countries, such as Romania, are preparing themselves for accession and are reorganizing their research and phytosanitary organizations. Other European countries just outside the Union such as Russia and Ukraine are rapidly adapting to consumer driven potato requirements with new systems of breeding, seed potato production and supply chains being established.

Many papers reflect societal and commercial changes such as consumer behaviour and marketing aspects in relation to fresh and processed potatoes in western, central and eastern Europe. Seed trade between western and central Europe is entering a new stage with altered inspection requirements, quarantine regulations and variety needs and acceptance. A new balance is needed to allow the high level of resistance to diseases in present varieties along with their current inherent sustainability in central and eastern Europe and the need to produce varieties that are accepted by consumers and processors. Increasingly, retailers and processors, Europe wide, lead the way at the global level to establish labels and certificates to assure food safety and sustainability. Agronomy research is aimed at improving the use of resources by a proper management of fertilization and disease control helped by the availability of precision farming techniques.

Various papers treat modern breeding techniques to improve quality – *e.g.* for technical purposes - but also to enhance resistance to diseases. Late blight caused by *Phytophthora infestans* receives special attention. Its resistance may be enhanced by a major effort to stack major genes from wild potato species in current varieties.

The last section of the book treats the developments in the potato supply chain, marketing and research and developments in a selected number of European countries: Germany, Poland, Hungary, the Czech Republic, Romania, Estonia and Russia. The diverse set of countries regarding the potato industry show how each deals with the socio-economic and environmental situation in varying degrees of development to adapt to a changing Europe.

Norbert U. Haase
Anton J. Haverkort
(editors)

Introduction

Challenges and opportunities of tomorrow's agricultural world

C.-A. Bartmer
German Agricultural Society (DLG), Eschborner Landstr. 122, D-60489 Frankfurt, Germany

The International Potato Congress and Potato Europe are good examples of the international approach to a theme for which national discussions and thinking patterns have become too constrained in view of liberalised markets. In addition to its international orientation, the Congress is characterised above all by the comprehensive approach of the technical themes selected that permeate the entire value chain. Thus, challenges and opportunities of tomorrow's agricultural world have to be taken into consideration, because they will lastingly influence the development of the entire sector, including the production of potatoes, that can play an increasingly more important role tomorrow.

- By comparison with today, the world population will have grown by 40 percent to some nine billion people by the year 2050. Experts estimate that twice as much food must be produced in the next 20 – 30 years as in the last 10,000 years together. In the last two years the world has brought in the highest grain harvests ever achieved, and at the same time, in view of growing demand, reduced stocks at 18 percent of annual consumption at the end of the grain business year 2005/2006 to the lowest level ever recorded.
- At present crude oil prices of over $ 70 per barrel are being obtained. This means that heating oil prices are above 60 ct/ltr. Accordingly, the thermal value of wheat today at € 24/100 kg is more than twice as high as the local miller is willing to pay for quality wheat. Biomass for energy production, as a fuel or as an industrial raw material, is beginning to account for a growing extent of cultivation and is competing with food production for arable and grassland area.
- This increased demand cannot be absorbed by expanding agriculturally used areas. On the contrary, every year some seven million hectares of agriculturally used land are irrecoverably lost by settlement, erosion, salinification and sanding.
- A further challenge in many regions of the world is the scarcity of water. It often represents the most important minimum factor for creating yield. According to the FAO, today 70 percent of the water volume is used for irrigation. The rest, in other words only 30 percent, is used for private consumption and industry. A forecast by the World Bank shows that in the year 2050 only 58 percent of the population will have sufficient water. The water scarcity will have two consequences: firstly, the volume of available irrigation water per hectare will decline, and secondly many areas will drop out of the irrigation system completely due to the higher priority of water use for other purposes. In this connection I don't need to stress how important irrigation systems are in potato production.
- Globally, on average more than 40 percent of harvests will be destroyed by weeds, diseases and pests.
- The liberalisation of markets will continue to advance. The trade policy interests articulated in the WTO negotiations allow increasingly fewer sectoral economic areas that hide behind customs barriers and export subsidies.

- Ever fewer financial resources will be available in public budgets in future to moderate the consequences of global competition for agriculture, that potato producers too have met with the new agricultural reform. As unfair as Blair's arguments on the financial comparison of science and research – that are chiefly funded nationally - by comparison with the "fully EU-integrated' old agricultural economy may be, they are despite this an "indicator" that at the latest by 2014 rededication of major funds can be expected under the "Lisbon Process" concept in order to enhance the competitiveness of the EU.
- In Europe there are opportunities for entrepreneurs. 450 million consumers with purchasing power on our doorstep represent the world's largest market. These consumers should be kept in mind when planning farm concepts. Consumers have precise ideas of tailored production of foods that they want to be safe, high quality, and as cheap as possible. There should be no complaints about these aspects, since this wish is very rational and is indeed practiced by every successful entrepreneur. However, consumers also have precise ideas about food production with rather fuzzy knowledge about the production process, the technology used, and the structuring of landscapes. Embodied in regulations from "good technical practice" to the agricultural non-word of the year, "Cross Compliance", farming entrepreneurs and agricultural administrations experience flourishing bureaucracy on the one hand, while on the other hand farmers are increasingly paying the price of an EU promotion system that makes access to new technologies more difficult for them.

Already today the initial effects of subsequent scarcity conditions can be felt on the world market through the rising raw material prices. For example, the world market price for sugar has more than doubled since 2003.

What agricultural products is tomorrow's consumer looking for? This is one of the most important questions that needs to be tackled repeatedly.

First of all, of course, foods, despite all the euphoria surrounding bio-energy, it is not possible to substitute foods. That is why the consumer will also be willing to pay a higher price for this. These foods should have high nutrient value. Varied and healthy ingredients are a special challenge to breeding. Simple preparation and enjoyment that – especially for potatoes - also includes aspects of provenance and production techniques in farming and food processing are important. But all this is nothing if the value chain cannot guarantee quality and safety.

Development of the sector cannot be driven by politics in the long term, but only by consistent attention to technical progress. This includes progress in breeding, in upstream industrial inputs, in improved mastery of processes, and in optimised farm organisation.

The times of incalculable agricultural surpluses presumably belong to the past. A new seriousness in the discussion on new technologies in farming is needed. These technologies are more than just progress in the technical sense. Willingness to innovation is a basic social attitude that finds its expression in courage to change. Progress in practice and research is the crucial key to securing affluence – in the European high-yield location, and especially in rural areas.

Curriculum vitae – Carl-Albrecht Bartmer

Carl-Albrecht Bartmer of Löbnitz a.d. Bode (Saxony-Anhalt) owns and manages a 1000 hectare farm. During the last 15 years, Bartmer has shown extraordinary pioneering spirit in turning around his family business into a successful enterprise, which dates back to 1735. Bartmer, 44, comes from a family in which entrepreneurial spirit and active collaboration with the German Agricultural Society are traditions. Following A-levels and an agricultural apprenticeship as well as academic agricultural studies in Goettingen, Germany, he was offered the opportunity to manage a seed-growing business in Schleswig-Holstein, North Germany. Further education in business management together with an international outlook are Bartmer's corner-stones in today's business environment. Since January 2006 Bartmer has been president of the German Agricultural Society (Deutsche Landwirtschaft-Gesellschaft/DLG).

Company profile – German Agricultural Society (DLG)

The German Agricultural Society (DLG), based in Frankfurt, Germany was founded in 1885 by Max Eyth. It pursues the goal of translating scientific findings into practice in the agricultural and food sectors. With over 18,000 German and international members, the DLG is a non-profit, self-help organisation, specialising in exhibitions, publishing, certified agricultural machinery and food testing. Serving the European and world interests at large, the society is one of the leading food and agricultural organisations in Europe. With its international activities and initiatives the DLG sets standards and provides impetus for progress.

The product and the association in transition – the mission of UNIKA

W. Hilse
Union der Deutschen Kartoffelwirtschaft e.V. (UNIKA, German Potato Industry Association), Claire-Waldoff-Strasse 7, 10117 Berlin, Germany

Abstract

Viable trade association work is increasingly oriented beyond the borders of Europe. The expansion of the European Union on May 1st, 2004 reinforced the trend of looking beyond products and markets, to take the stakeholders and their organizations into account and to explore mutual interdependencies. For innovative trade associations, service offerings for members come first and foremost.

Keywords: service offerings, value chain, image, internationalization

Introduction

The "European Potato World" is meeting for the first time ever, in Germany and the federal state of Lower Saxony, on the occasion of the PotatoEurope Congress 2006.

The initiator of the Congress, the *Union der Deutschen Kartoffelwirtschaft e.V.* (UNIKA, German Potato Industry Association), and the organizer of the Congress, the *Deutsche Landwirtschaftsgesellschaft e.V.* (DLG, German Agricultural Society), picked up on the idea for such an event, which originated in the Netherlands and France, and turned it into reality. The participants have already decided that UNIKA and the DLG will host the Congress in Germany again in 2010 after Belgium, France, and the Netherlands take turns in the intervening years.

Market influences and the work of the trade associations

On May 1st, 2004, Germany became the geographic center of Europe. The expansion of the European Union by ten states in central and eastern Europe is both a challenge and an opportunity for an association like UNIKA and all the companies that participate in the potato value chain. This is a factor that will determine success or failure on the markets. The main objective of companies in this industry, however, remains adding value and earning money thereby.

What role do the trade associations play in such a period of upheaval? Does their mission change? How can we describe the repercussions for members?

The potato, as a product, has always been at the mercy of the free market. Markets don't operate according to the principles of justice; they are unconditionally linked to supply and demand. The parameters of innovation, quality, and price are the forces driving the market. The challenge here is taking leadership. The markets influence, demand, and shape the active stakeholders in the entire potato value chain - from growers to traders or processors.

UNIKA and its mission

On October 30th, 2001, in Hamburg, the founding associations – *Deutscher Bauernverband* (German Farmers' Association), *Bundesverband Deutscher Pflanzenzüchter* (German Plant Breeders Association) – succeeded in creating UNIKA, after two predecessor organizations. As a innovative and future-oriented association, its mission is to represent the "entire world of the potato", enhance the perceived value of the product while increasing its actual value, and provide services to its members as an entrepreneurial trade association. To a large extent, UNIKA's role is to coordinate and bundle the knowledge of experts who volunteer their time to the association and make this information available to its members and other institutions.

Last but not least, its mission includes bringing together people and organizations whose companies compete with one another in the course of daily business, and moderating a discourse between them. The Dutch and the French have shown us how an industry can speak with one voice in all regions and at all levels. This appears to be crucial not only in a unified Europe, but also given the rapid internationalization and globalization of agribusiness. In fact, we might ask - is there not a pressing need for the European potato industry to speak with one voice?

The potato is a modern product with a "trendy" character that hopes to appeal to the current health fad and enjoyment wave. The image of the spud is also influenced by other parties that are involved on the product's road to the market, as well as by the trade associations and other opinion-makers. As such, UNIKA is the nexus for communication and information and serves as an organizational center. In this role, the association's so-called "social expertise" takes on a surprising significance in its daily operation as a service organization.

Trade associations need their members to identify with them

The extent to which members identify with their organization is crucial for the survival of a trade association. This cannot be achieved through a functioning office alone. An influential corps of honorary officials that is ready to take action also plays a decisive role. Work in expert commissions or other subsidiary bodies is essential in that it allows associations to be filled with life. The expert knowledge of the business people and specialists in the entire industry should be better integrated in the future, so that the positive feeling of belonging to a sector with a bright future does not remain a mere slogan. The EU expansion to the East has also reinforced the atmosphere of excitement in the potato industry.

Modern guiding principles

In companies of all types, discussions have moved beyond the familiar topics of small or large, slow or fast, and have shifted to include the concepts of cooperation, innovation, and the value chain. The expression "positive realism" aptly describes the current challenge in the market. As the mouthpiece of the German potato industry, UNIKA has competitiveness, increasing sales, and the image of the potato clearly in its sights. In close coordination with its partners, Germany must continue to maintain its position as a "potato country" in the future.

Conclusion

Successful work has always been the result of cooperation among dedicated people in an organized and transparent group. This is the platform that UNIKA offers for breeders, producers, marketers, and processors, with the promise that it will do everything in its power to ensure the future of its members.

Curriculum vitae – Werner Hilse

Werner Hilse was born on March 26th, 1952 in Schnega/Warpke, Lüchow-Dannenberg in Lower Saxony in Germany. He is married and has three children.

He completed the vocational training course to become a *staatlich geprüfter Landwirt* and *Landwirtschaftsmeister* (state-certified farmer and agricultural master craftsman). At the age of 18 he took over his parent's farm, from then on working mostly on his own responsibility. Today it is a 300-hectare farm growing corn and grain, potatoes, sugar beets, and a hog-raising facility (with stalls 1,000 animals), as well as a turkey farm (with stalls 40,000 animals) operated in cooperation with other partners. In 1981, he became volunteer chair of the *Landvolkbezirksverband* (farmers' district association) of Schnega-Bergen. In 1984 he was elected as member of the *Kammerversammlung der Landwirtschaftskammer* (Joint Assembly of the Chamber of Agriculture) of Hannover.

He was the Vice President of the *Landesbauernverband* (State Farmers' Association) of Lower Saxony from 1990 to 2003 and has been its President since 2003. He has worked in the representative bodies of the global starch company AVEBE since 1992 and is the deputy chairman of its advisory board. He serves as chairman of the CESPU (Committee of the European Starch Potato Producer's Union) in Brussels. His work on behalf of the potato industry includes his activities as the chairman of the potato committee of the *Deutscher Bauernverband* (DBV, German Farmer's Association), and as chairman of the *Bundesverband der Stärkekartoffelerzeuger* (BVS, German Association of Starch Potato Producers), and chairman of the *Union der Deutschen Kartoffelwirtschaft* (UNIKA, German Potato Industry Association). He is Chairman of the *Centrale Marketing Gesellschaft der Deutschen Agrarwirtschaft* (CMA, Central Marketing Organization of German Agricultural Industries). Since June of this year, is vice president of the *Deutscher Bauernverband*.

Company profile – UNIKA

The *Union der Deutschen Kartoffelwirtschaft e.V.* (UNIKA, German Potato Industry Association), founded in 2001 in Hamburg, was created to increase the competitiveness of German potato growers, breeders, marketersand processors. As the interest group of the German potato industry, it focuses the concerns of the sector and lobbies the political and economic conditions affecting the potato sector. Its headquarters are in the 'House of Agriculture and Food Industries' in Berlin, where it serves its 57 members. The Union is headed by its volunteer Board of Directors with support from the business office.

Expert commissions, comprising specialists from specific areas of the potato industry, ensure that the business office and the trade association perform their task based on expert professional knowledge. An Advisory Council, also comprising volunteers, provides stimulus and suggestions for the strategic orientation of the association and the establishment of expert commissions by the Board and the Management.

Modern breeding practices

Overview of biotechnological breeding possibilities

Gerhard Wenzel and Barbara Ros
Chair of Plant Breeding, Technical University of Munich, D-85350 Freising-Weihenstephan, Germany

Abstract

Agronomy and agrochemistry have reached a level where further progress is neutral in relation to yield. Scientific progress helps environmental needs. Genetics alone may close the yield gap by a combined application of classical techniques and molecular knowledge. Large-scale genome analysis and related technologies provide access to a refined understanding of the genome. Fortunately, molecular markers are a genetic tool accepted by the public which often judges gene technology critically. There is a strong tendency that classical combination breeding together with marker assisted selection is more effective in achieving complex breeding goals than gene transfer approaches. Here, tools will be elucidated such as higher efficiency in parent selection; controlled combination of heterocygous parents; uncovering a rare but desired genotype in a large segregating population; or pyramiding single traits to result in more complex characters, *e.g.* durable resistance. Such techniques allow more reproducible and faster development of new and better cultivars. Today a breeder has to know how such modern breeding tools work and how to use them in an intelligent manner.

Keywords: biotechnology, classical breeding, genetic markers, genomics

Introduction

Due to its tetraploid nature and its high levels of self incompatibility, potato improvement is not an easy task with classical genetic tools, but it is the first crop plant to which biotechnological methods were successfully combined and applied. Thus, it will become possible to construct key strategies fitting breeder's dreams, like:

- secure germ plasm conservation;
- handling a crop plant at different ploidy levels;
- characterization of traits in large populations;
- programmed combination of such traits;
- knowledge directed construction of metabolic pathways.

Of course, classical and such new approaches should be combined in an intelligent way. Whenever classical techniques work, biotechnology has only a chance when it is more efficient or cheaper. Figure 1 gives an example how the two methods complement each other.

A central change in breeding strategies is the conversion from phenotypic evaluations to genotypic ones. Since here molecular markers are the key tool, their principle will be discussed first, before their application is described in the two breeding areas: (a) maintenance of genotypes and (b) breeding of new cultivars.

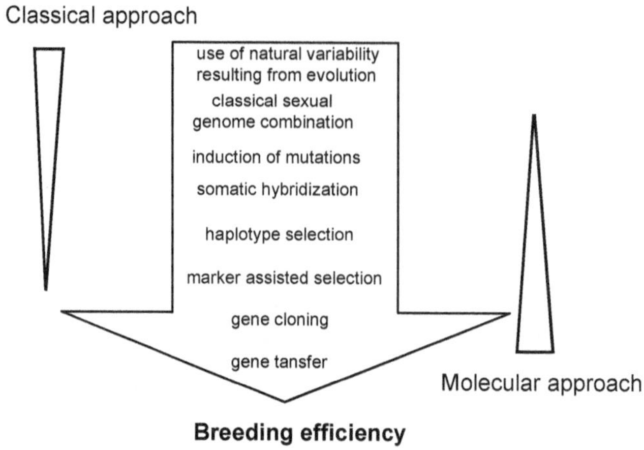

Figure 1. Considerations for using biotechnology in potato breeding.

Markers as tools for evaluation of stability and variability

In Mendelian breeding programs phenotypic characteristics are used to distinguish between a good and the very best clone. Since simple segregating morphological markers are not too numerous, it was an advantage when biochemical and molecular markers were developed. Today, the most powerful markers are molecular ones since they are unlimited in number, easy to generate, independent from developmental stages and without epigenetic effects. The first system developed was the restriction fragment length polymorphism (RFLP, Botstein *et al.*, 1980). Marker alleles are identified by size differences of the restriction fragments to which the probes hybridise and correlated to similarly segregating phenotypic traits of the mapping population. However, RFLP analysis is a time-consuming and labour-intensive task and, therefore, not suitable for the rapid genotyping of large populations. Here, random amplified polymorphic DNA (RAPD, Williams *et al.*, 1990) found its use. Addition of a 10-base oligomer to restricted DNA results in RAPDs which can be directly visualized after PCR amplification in gel electrophoresis. This has striking advantages such as low costs, the small amount of DNA needed (10-100 ng), and the use of universal primers. However, the DNA profiles are often non-reproducible. Thus, amplified fragment length polymorphism (AFLP, Vos *et al.*, 1995) is presently the most efficient technique for detecting polymorphisms in potato. AFLP generates many bands, but not too many to create difficulties in scoring. Another technique used quite often for potatoes, is simple sequence repeats (SSRs) or microsatellites (Tautz and Renz, 1984). Since SSR loci are evenly spaced throughout the genome as simple PCR markers, they represent an ideal resource for a marker class. The next generation of genetic markers is single nucleotide polymorphism (Reiter, 2001). A SNP is a single nucleotide position in a defined DNA stretch at which there is variation between different clones. SNPs are of particular interest for potato improvement, since they represent the most frequent variations in the genome, and can be simply treated as di-allelic markers. The most advanced SNP procedures are no longer gel based and, therefore, readily automated for high throughput genotyping, allowing cost-effective screening of large populations.

In the tetraploid potato all markers face the problem that four alleles exist per locus, and most markers show only the presence or absence but not the quantity of an allele. So it is difficult to distinguish a simplex from a triplex situation.

Cytoplasmic genomics have also been included for characterization. Plastid (pt)DNA is less variable compared to mitochondrial DNA and contains predominantly two pt-types: tuberosum (T)-type and wild (W) type chloroplasts in a 1 : 1 ratio. Provan *et al.* (1999) emphasize the need to increase levels of non-T-type cytoplasms in the cultivated potato gene pool, since most promising new traits are coming form wild species and primitive cultivars carrying W type pastids used as females and thus diluting the T-type substantially.

Whether molecular markers should be also used in seed boards for legal variety characterization is under discussion. As long as most markers are only correlated to traits, often located in introns rather than in exons or functional genes, marker profile variation is no proof for a deviation in a registered character. Thus, varieties may easily express 5% variation in their AFLP profile during successive years without any morphological change. This consideration counts also when deciding whether two clones in a gene bank are duplicates or not.

Techniques for genotype maintenance

Potato breeders have to maintain valuable accessions and rare allele combinations. Thus, *in vitro* living collections were established, from which about 3 million potato plantlets are rapidly propagated (Preil, 2003). By performing molecular genetic analysis of such collections new challenges are opened up as *e.g.* resolving questions of taxonomy or examining the population biology of breeding systems and ploidy levels (Bryan *et al.*, 2005).

Protoplasts offer in theory a much faster propagation of a given genotype since a single potato leaf may contain several hundred thousand cells, each of which may grow into a functional plant. Although numerous protocols exist for an efficient protoplast regeneration of potato through a callus phase, this technique is not used routinely, since a crucial prerequisite of rapid propagation - genome stability - is not fulfilled.

Potatoes can be stored in liquid nitrogen, to avoid the laborious transfer process in living collections, and to make sure that no undesired somaclonal variation influences a valuable genotype. Such techniques are of particular interest in maintaining germ plasm collections. In the German potato collection, *e.g.*, around 1000 potato clones are cryopreserved (Keller, 2003). To avoid any callus growth, embryos or embryoids are cryopreserved and plantlets are regenerated via embryoids.

All information gathered during maintenance, particularly those on genetic distances, is a useful prerequisite to select optimal parents for sexual and somatic combinations when breeding new cultivars.

Breeding new cultivars

Increased information of structure, function, and regulation of the potato genome helps in the two phases of plant breeding: the production and use of variability as well as in the selection phase. Biotechnology offers the following opportunities to speed up and increase efficiency of this processes:

- Working at the mono- and/or dihaploid ploidy level;
- Using marker assisted selection;
- Transfer of genes either via symmetric and asymmetric fusion, or by DNA transfer.

Since the 60s breeding at the dihaploid (2n = 2x = 24) level is an established procedure. Adequate percentages of dihaploid potatoes can be produced parthenogenetically (Hougas and Peloquin, 1957), a technique which worked as well for the reduction of the ploidy level from 2x to monohaploids (2n = 1x = 12; van Breukelen *et al.*, 1975). Dihaploids are of particular interest when combinations of diploid species with *S. tuberosum* are an aim.

The hart of Mendelian genetics, sexual combination, is always accompanied by meiosis and recombination, and thus its specific success is unpredictable even in dihaploids. Fusion of protoplasts allows to combine asexually two complex nuclear genomes and to combine plasmones as well. Presently, the electrofusion technique is applied in most laboratories with a lamellar fusion chamber. Hybrid identification is predominantly performed by molecular probes which allow the use of undifferentiated calli, so that waiting for morphogenesis is not necessary. In experiments with symmetric protoplast fusion Möllers *et al.* (1994) could demonstrate a substantial yield increase by comparing the mean tuber yield of the parents with the tuber yield of the hybrids. There were, however, large variations between different combinations. The fusion of primitive cultivars or wild species with *S. tuberosum* concentrates on the integration of characters not available in a heritable form in the primary gene pool of *S. tuberosum* (Table 1). For distant hybridisations, asymmetric fusion is better suited than normal symmetric fusion. Molecular strategies for selection are useful when the agronomically important characters are not coded just by one gene but by several ones. McGrath *et al.* (2002) detected, *e.g.* one clone with 48 chromosomes carrying a single *S. brevidens* marker, suggesting transfer of just this *S. brevidens* DNA piece into *S. tuberosum*. Such hybridisation techniques come very close to gene transfer approaches, but final trait expression appears in a less-predictable process.

Asymmetric hybrids with unequal contributions of the cytoplasmic and nuclear genetic material of each parent are sometimes of greater interest than symmetric fusions. In the simplest procedure, protoplasts of a donor species (normally a distantly related genotype) are irradiated, destroying the nuclear DNA before fusion with the recipient protoplast (normally a cultivar). As a result, the nuclear genome of the cultivar is combined with the plastome of a distantly related species.

Somatic cell fusion, circumvents the normal uni-parental inheritance of the plasmone, and permits complete combination of cytoplasms. The interaction of these compartments has influence on several yield traits. Evaluations of the correlation between the cytoplasmic types showed *e.g.* that an interaction existed between starch content and mt-pt combinations (Lössl *et al.*, 2000). It is assumed that an advantage of clones with an optimized organelle

Table 1. Selected characters combined via somatic fusion (for literature see Wenzel 2006).

Parents	Characters of interest
Symmetric fusion	
S. tuberosum (+) *S. tuberosum*	PVX/PVY
S. tuberosum (+) *S. tuberosum*	Nematode resistance
S. tuberosum (+) *S. brevidens*	PLRV resistance
S. tuberosum (+) *S. brevidens*	PVX resistance
S. tuberosum (+) *S. brevidens*	PVY resistance
S. tuberosum (+) *S. brevidens*	*Erwinia* resistance
S. tuberosum (+) *S. bulbocastanum*	*Phytophthora* resistance
S. tuberosum (+) *S. bulbocastanum*	*Meloidogyne* resistance
S. tuberosum (+) *S. commersonii*	*Erwinia* resistance
S. tuberosum (+) *S. commersonii*	Frost tolerance
S. tuberosum (+) *S. tarijense*	*Erwinia* resistance
S. tuberosum (+) *S. verrucosum*	PLRV resistance
Asymmetric fusion	
S. tuberosum (+) *S. commersonii*	Verticillium resistance
S. tuberosum (+) *S. circaeifolium*	
S. tuberosum (+) *Lycopersicon esculentum*	Chilling tollerance

composition exists and that they regenerate *in vitro* preferentially (Frei *et al.*, 1998). The effect was, however, not large enough to significantly produce in large scale field trials more starch, better yield or superior morphological characteristics (Braun *et al.*, 2002).

Different *in vitro* approaches can be combined. Wenzel *et al.* (1979) proposed the combination of haploidy, protoplast fusion and classical breeding steps for combining several traits of potato in a stepwise fashion. First the monohaploid ploidy level is reached via two successive haploidization steps. Homozygous dihaploids are subsequently obtained by doubling of 1x clones. In these clones additive traits are homozygous and will be inherited during a sexual combination to heterozygous interdihaploids. These interdihaploids are subjected to intensive sexual combination and selection to give superior heterozygous interdihaploids. The final step is the somatic fusion of these hybrids to combine several characters and to reach, at the same time, the tetraploid level expressing the highest amount of heterozygosity. About 15 years after this scheme had been proposed first analytical synthetic clones combining different resistances were tested in the field (Thach *et al.*, 1993).

Application of molecular markers in breeding of new cultivars

Application of modern breeding techniques relies to a great extent on strategies using available possibilities rather than on a programmed approach solving problems in a logical way. The knowledge of physiological pathways and cascades is still too limited to effectively manipulate specific enzymes. The main activities today are characterized by efforts to construct comprehensive genomic maps. In the next phase, research activities will focus on the utilization of these maps for gene cloning and verifying the function of the responsible genes. A systematic molecular evaluation of the complete genetic information and the resulting cellular activities, such as transcription and protein expression will be essential for future developments. The increasing amount of information about DNA documented in molecular marker collections and dense gene maps, together with an excellent bioinformation system, allows the development of new strategies for creation of superior cultivars. Common principles are uncovered, *e.g.* in defence reactions to biotic and abiotic stress, offering strategies like the candidate gene approach (Lübberstedt *et al.*, 2002). Molecular markers are a powerful tool for: a) higher efficiency in parental selection allowing a controlled combination of better combining and thus heterotic parents and b) pyramiding of single traits to result in more complex characters (Ranjekar *et al.*, 2002). Additionally, they are presently the most efficient system to uncover a rare but desired genotype in a large, segregating population.

Most mapping and marker isolation has been done at the diploid level for monogenic traits, since in the tetraploid potato the dosage effect by doubling the ploidy level might have striking influence. A huge number of agronomically important genes have been correlated to molecular markers, but only some of them are gene-sequence derived and only these do really co-segregate with the target gene.

Of utmost interest are alleles responsible for resistance to diseases and pests. Understanding the genetic basis of pathogen resistance has been broadened in potato tremendously thanks to DNA-marker technologies. Here, resistance gene analogs (RGAs) become more and more interesting as molecular markers. RGAs are primarily located close to resistance genes and are often members of multigene families (Lübberstedt *et al.*, 2002). Markers based on the use of RGA approaches often have more general diagnostic value. Since in potato hot spots with numerous resistance loci exist, such areas are often associated with the relevant characters over a wide range of genotypes (Gebhardt *et al.*, 2005).

Marker assisted selection (MAS) offers the opportunity to combine different genes for a given pathosystem in a single genotype (gene pyramiding). While many important genes are monogenic, most characters relevant for the tuber crop show continuous phenotypic variation because they are controlled quantitatively by multiple genes (QTL) and by environmental factors. The situation for such traits is less satisfactory. Tests were started to understand which candidate gene loci are causal for the QTL tuber starch, yield and chipping quality. Even after molecular localization of a QTL it is an unsolved question how to apply this knowledge in breeding programs.

Many genes have been introgressed into potato cultivars and breeding clones from other tuber-bearing *Solanum* species by sexual and somatic hybridisation. Such clones are used for constructing molecular maps with different sets of RFLP, RAPD, AFLP and SSR markers. Molecular potato maps are now available (Meyer *et al.*, 2005) and a physical map is under construction (Bachem *et al.*, 2005). The molecular maps revealed that tuber starch content, sugar content and yield are linked on chromosome V. This same genomic region also harbors QTL for plant maturity and vigor (Visker *et al.*, 2003). Furthermore, some genes functional in carbohydrate metabolism or transport are closely linked to QTL for tuber traits (Menendez *et al.*, 2002).

Table 2. Some important and mapped DNA markers (for literature see Wenzel 2006).

Trait	Gene	Chromosome
Potato virus Y	Ry_{adg}	XI
	Ny_{tbr}	IV
Potato virus Xa	*Rx1; Rx2*	XII; V
	Na	XI
	Nb	V
	Nx	IX
Potato leaf roll virus	*PLRV* QTL	XI
Globodera rostochiensis	*Gro*	III; VII;X;XI
	H1	V
Globodera pallida	*Gpa* QTL	IV;V;IX;XII
	Gpa2	XII
Phytophthora infestans	*RB*	X
	R1	V
	R2	IV
	R3; R6; R7	XI
	R_{blc}	VIII
	QTL	V
Synchytrium endobioticum	*Sen1*	XI
Erwinia carotovora	QTL	
Tuber starch; tuber yield	QTL	I-XII
Cold sweetening	QTL	V; IX
Skin colour	QTL	X
Tuber flesh color	QTL	III

Such clusters make further map based cloning more promising, since sequencing strategies can concentrate on particular DNA fragments.

Mapping of traits is a prerequisite for gene cloning by chromosome walking. For map-based cloning ultra dense maps with distances of less than 0.1 cm (1000 kb) between trait and marker have to be available. Furthermore, BAC libraries covering the whole genome two to four times are needed. Since these are very heavy prerequisites, cloning of genes where the gene products are unknown, is a most difficult task.

Gene transfer

In addition to the techniques already discussed, the transfer of cloned genes allows improvement of a superior cultivar in just one or a few missing traits. Prerequisites are:

- availability of the trait to be transferred as cloned DNA;
- availability of a powerful transfer system;
- availability of a reliable regeneration system.

For potato, the last bullet point is guaranteed by single cell regeneration. From the two transfer systems: micro bombardment and vector assisted transfer via the *Agrobacterium* system are the two transfer systems of choice. Rather general is the use of *Agrobacterium* where large fragments can be transferred. Transfer of monogenic characters and even the transfer of more than two genes is becoming routine. New vectors have been developed composed entirely of potato DNA, which are already capable or replicating in bacteria (Lokerse *et al.*, 2005). Despite such progress it is still an unsolved question how to address complex traits. Since most traits of interest are not coded by a single gene but rather by oligo- or polygenic ones, different strategies have to be applied for complex characters. Presently MAS is allowing faster success than the transfer of isolated or even monogenic genes.

Several characters have been transferred today using micro-bombardment and *Agrobacterium* systems (Table 3). Of particular interest are genes responsible for the carbohydrate metabolism (Chen *et al.*, 2001) and for disease resistances (Gebhardt and Valkonen, 2001).

Clues of the biological function of genes can be obtained from their spatial and temporal expression patterns. Expressed sequence tags (EST) from different tissues provide preliminary information on expression patterns for the more abundant transcripts. EST databases turned out to be excellent resources for identifying genes by sequence comparison, for discovering new genes, and for assigning exons in genomic sequences.

Ros *et al.* (2005) used such techniques to search for genes, which are induced during the infection of potato with *Phytophthora infestans*. Leaflets of the tolerant cultivar Bettina and the susceptible cultivar Indira were inoculated with *P. infestans*. Using infected and control tissue, two libraries were prepared, highly enriched for *P. infestans* induced genes. In such libraries nearly 300 unigenes were found, and 182 cDNA clones were selected for further analysis of the differential gene expression. This allowed to access a source of *P. infestans* induced genes, like pathogenesis related protein 1, lipid saturase, peroxidase, chitinase, glucanase, with increased activity and others like proteinase inhibitor, leucin aminopeptidase, or RNA binding protein being reduced. It appears that defence reactions need energy; taking this energy from the plant growth metabolismus. Since resistant potato cultivars are not less yielding, a breeder always selects intuitively for both: increased resistance and increased growth metabolism.

Table 3. Selection of traits expressed after gene transfer (for literature see Wenzel 2006).

Trait	Trait
PVX resistance	Palatinose production
PVY resistance	Zeaxanthin production
PLRV resistance	*Rb; R1* resistance gene against *Phytophthora*
Bt Tolerance	Pectate lyase
Herbicide tolerance	*Erwinia* resistance
Inulin production	Starch synthase
Increased level of phytochrome B	Branching enzyme
Cyclodextrine production	Vaccine production

Conclusions

The number of breeding aims is steadily increasing. Thus, the addition of biotechnological methods to classical breeding approaches increasing the efficiency in breeding programs is crucial today. Most universal is the use of tissue culture for rapid propagation. It should be stressed, however, that biotechnology alone will not make a cultivar. One needs Mendelian genetics, classical clonal selection and intensive field testing for an economic combination of around 40,000 genes in a superior clone. Such selections can be backed by MAS and in rare cases a single trait may be altered by gene transfer. And even the most superior clone produced by intelligent combination of such approaches will not remain superior. The dynamics of evolution will make breeding of new cultivars an indefinite task. All techniques are needed allowing to solve old problems like *Phytophthora* resistance and new ones in consequence of potato becoming a renewable resource crop.

References

Bachem, C. and R. Visser (2005). Update on the status of the potato genom-IC sequencing consortium (PGSC). 2. Solanaceae Genome Workshop, Ischia, Italy. p. 109.

Botstein, D., R.L. White, M. Skolnick, and R.W. Davis (1980). Construction of a genetic linkage map in man using restriction fragment length polymorphisms. American Journal of Human Genetics 32:314-331.

Braun, A., C. Gatz, U. Frei, and G. Wenzel (2002). Field experiments with transgenic potato clones. In: Wenzel G, Wulfert I (eds), Potatoes Today and Tomorrow. p 210, EAPR.

Bryan, G., G. Ramsay, K. McLean, N. Massat, R. Waugh, and D. Spooner, (2005). Molecular genetic analysis of a large potato genebank. 2. Solanaceae Genome Workshop, Ischia, Italy. p. 89.

Chen, X., F. Salamini, and C. Gebhardt (2001). A potato molecular function map for carbo-hydrate metabolism and transport. Theoretical and Applied Genetics 102:284-295.

Frei, U., M. Stattmann, A. Lössl, and G. Wenzel (1998). Aspects of fusion combining ability of dihaploid Solanum tuberosum L. influence of the cytoplasm. Potato Research 41:155-162.

Gebhardt, C., A. Ballvora, and E. Ilarionova, (2005). Genomic sequencing and functional characterization of a resistance hot spot on potato chromosome V. 2. Solanaceae Genome Workshop, Ischia, Italy. p. 108.

Gebhardt, C. and J.P.T. Valkonen (2001). Organization of genes controlling disease resistance in the potato genome. Annual Review Phytopathology 39:79-102.

Hougas, R.W. and S.J. Peloquin (1957). A haploid plant of the potato variety Katahdin. Nature 180:1202-1210.

Keller, J.E.R. (2003). In-vitro-Kultur zur Erhaltung pflanzengenetischer Ressourcen. pp. 51- 52, ADVK Meeting, Sept. 10-12, Geisenheim, Germany.

Lokerse A., A. Erasmuson, S. Baldwin, P. Cooper, J. Jacobs and A. Conner (2005). An intragenic vector for gene transfer to potato without foreign DNA. 2. Solanaceae Genome Workshop, Ischia, Italy. p. 118.

Lössl, A., M. Götz, A. Braun and G. Wenzel (2000). Molecular markers for cytoplasm in potato: male sterility and contribution of different plastid-mitochondrial configurations to starch production. Euphytica 116:221-230.

Lübberstedt, T., V. Mohler, and G. Wenzel (2002). Function of genetic material: Genes involved in quantitative and qualitative resistance. Progress in Botany 63:80-105.

McGrath, J.M., C.E. Williams, G.T. Haberlach, S.M. Wielgus, T.F. Uchytil, and J.P. Hegeson (2002).Introgression and stabilization of *Erwinia* tuber soft rot resistance into potato after somatic hybridisation of Solanum tuberosum and S. brevidens. American Journal of Potato Research 79:19-24.

Menendez, C.M., E. Ritter, R. Schäfer-Pregl, B. Walkemeier, A. Kalde, F. Salamini, and C. Gebhardt (2002). Cold-sweetening in diploid potato. Mapping QTL and candidate genes. Genetics 162:1423-1434.

Möllers, C. and G. Wenzel (1992). Somatic hybridization of dihaploid potato protoplasts as a tool for potato breeding. Botanica Acta 105:133-139.

Preil, W. (2003). In vitro Vermehrung gartenbaulicher Kulturpflanzen in Europa - aktueller Stand und Trends. pp. 51-52, ADVK Meeting, Sept. 10-12, Geisenheim, Germany.

Provan, J., W. Powell, H. Dewar, G. Bryan, G.C. Machray, and R. Waugh (1999). An extreme cytoplacmic bottleneck in the modern European cultivated potato (Solanum tuberosum) is not reflected in decreased levels of nuclear diversity. Proceedings Royal Society London B 266:633-639.

Ranjekar, P.K., A.P. Davierwala, and V.S. Gupta (2002). DNA Markers and heterosis. In: Jain SM, Brar DS, Ahloowalia (eds) Molecular techniques in crop improvement. pp. 161-201, Kluwer Academic Publishers, Dordrecht.

Reiter, R. (2001). PCR-based marker systems. In: Phillips R.L., Vasil I.K. (eds) DNA-based markers in plants, 2nd edition, Advances in Cellular and Molecular Biology of Plants, Vol. 6, pp 9-29, Kluwer Academic Publishers Dordrecht.

Ros, B., F. Thümmler and G. Wenzel (2005). Comparative analysis of Phytophthora infestans induced gene expression in potato cultivars with different levels of resistance. Plant Biology 7:686-693.

Tautz, D. and M. Renz (1984). Simple sequences are ubiquitous repetitive components of eukaryotic genomes. Nucleic Acids Research 12:4127-4138.

Thach, N.Q., U. Frei, and G. Wenzel (1993). Somatic fusion for combining virus resistance in Solanum tuberosum L. Theoretical and Applied Genetics 85:863-867.

van Breukelen, E.W., M.S. Ramanna, and J.G.T. Hermsen (1975). Monohaploids (n = x = 12) from autotetraploids Solanum tuberosum (2x = 4x = 48) through two successive cycles of female parthenogenesis. Euphytica 24:567-574.

Visker, M.H.P.W., L.C.P. Keizer, H.J. Van Eck, E. Jacobsen, L.T. Colon, and P.C. Struik (2003). Can the QTL for late blight resistance on potato chromosome 5 be attributed to foliage maturity type? Theoretical and Applied Genetics 106:317-325.

Vos, P., R. Hogers, R. Bleeker, M. Reijans, T. van de Lee, M. Hornes, A. Frijters, J. Pot, J. Peleman, M. Kupier, and M. Zabeau (1995). AFLP: a new technique for DNA fingerprinting. Nucleic Acids Research 23:4407-4414.

Wenzel, G., O. Schieder, T. Przewozny, S.K. Sopory, and G. Melchers (1979). Comparison of single cell culture derived Solanum tuberosum L. plants and a model for their application in breeding programs. Theoretical and Applied Genetics 55:49-55.

Wenzel, G. (2006). Biotechnology in potato improvement. In: Gopal, J. and Khurana, S.M.P. (eds.) Potato production, improvement and post-harvest management. pp. 109-146, Harworth, New York.

Williams, J.G.K., A.R.K. Kubelik, J.L. Livak, J.A. Rafalski, and S.V. Tingey (1990). DNA polymorphisms amplified by random primers are useful genetic markers. Nucleic Acids Research 18:6531-6535.

Curriculum vitae – Gerhard Wenzel

Gerhard Wenzel, born in Wuppertal, Germany (16.2.1943). Studied at the University of Köln. Promotion (1971) in botany, biochemistry, and genetics; Habilitation (1979) in botany. 1971-1972 Research assistant at the Max-Planck-Institut für Züchtungsforschung, Köln; projekt leader (1973-78), MPI für Pflanzengenetik, Ladenburg near Heidelberg; head of an independent research group (1979-81) MPI für Züchtungsforschung, Köln; director of the Institut für Resistenzgenetik (1981-96), Grünbach; since 10/93 TU München, Wissenschaftszentrum Weihenstephan, Ordinarius at the chair of Plant Breeding, D-85350 Freising, Tel.: 08161/713421; Fax: 08161/714511, E-mail: gwenzel@wzw.tum.de

Research work: Combined application of classical and biotechnological methods in breeding of crop plants with focus on barley, potato and wheat. Cell culture, molecular techniques and *in vitro* selections are applied. Aim is the improvement of resistances against biotic and abiotic stress, and improvement of quality. In addition, monitoring of transgenic rapeseed, maize, and potatoes is performed in field experiments.

Others: President of the German society for plant breeding.

Profile – Technical University of Munich: Chair of Plant Breeding

The chair of Plant Breeding is part of the department Plant Sciences at the Centre for Life and Food Sciences of the Technical University of Munich. Main research focus concentrates on production of healthy crops by classical breeding, *in vitro* approaches via haploids and somatic genetics, molecular genome and gene diagnoses, marker assisted selection including QTL, and functional analysis of genes for disease resistance. These methods have predominantly been applied on potato, barley, wheat, rapeseed, maize, and to a minor extent on rye, ground nuts, banana and apple. Presently research concentrates on complex characters like Phytophthora resistance in potato, Fusarium resistance in wheat or lignin content in maize. For wheat/ Fusarium a TILLING population is under construction, and for the lignin complex association mapping is used. Molecular techniques were also adopted in developing monitor programs by preformed chips to identify not only transgenics but also the respective cultivar. This is done particularly in rape seed. In the area of agronomy the focus was on the optimisation of nutrient uptake in cereals and potato. The area agronomy was separated from the breeding activities in 2004, so that the chair concentrates now on plant breeding only.

The chair of Plant Breeding is closely cooperating with four start-up companies founded by members of the institute: EpiLogic, EpiGene, Vertis Biotechnologie AG and GeneGlobal, all located at the university campus at Freising-Weihenstephan.

Modern methods for modern potato breeding programmes

M.F.B. Dale and J.E. Bradshaw
Scottish Crop Research Institute, Invergowrie, Dundee, DD2 5DA, Scotland, United Kingdom

Abstract

The potato industry in Britain and in Europe is trying to increase potato usage in an economically and environmentally sustainable way. New cultivars are required to give higher yield of saleable product at less cost of production and to meet new requirements by an increasingly demanding processing industry. Such new cultivars are required to have improved inherent resistance to pests and diseases, increased water and mineral use efficiency, and to meet consumer demands for healthy, flavoursome, convenience foods. Conventional potato breeding continues to involve making crosses between pairs of parents with complementary features and the selection of clones from within seedling progenies with as many desirable characteristics as possible for release as new cultivars. Increasingly, parents will have genes from wild species and come from complementary groups of germplasm to exploit hybrid vigour. Furthermore, as genetical knowledge accumulates, these parents will be chosen to have desired genes which will be selected in their offspring. The application of robust molecular markers to select important heritable traits promises to make selection both more efficient and also potentially more cost effective by reducing or replacing expensive and expansive trialling and testing procedures which are currently required, often over many sites and/or seasons. Successful cultivars may be genetically modified to introduce genes not present in cultivated potatoes and their wild relatives to achieve further desirable improvements.

Keywords: breeding, genetics, molecular marker, wild species

Introduction

The European cultivated potato, *Solanum tuberosum* subsp. *tuberosum,* was derived from a narrow genetic base of a few introductions of subsp. *andigena* and subsp. *tuberosum* from South America in the late 16th century, and possibly further casual introductions in the 17th and 18th centuries. As a consequence, it lacked sufficient genetical variation to afford adequate levels of resistance to a number of pests and pathogens which became a problem once it became a staple food crop in many countries. Most notable pathogens have been late blight (*Phytophthora infestans*) since the middle of the 19th century and the golden and white potato cyst nematodes (*Globodera rostochiensis* and *G. pallida*) in the 20th century. Hence, during the last 100 years, potato breeders have been introgressing genes for disease and pest resistance into Tuberosum potato varieties from the wild and cultivated species of Central and South America. However, the production of such new varieties using improved selection procedures does not guarantee commercial success. The two most widely grown cultivars in Britain are Maris Piper and Estima which were first recommended in 1966 and 1976, respectively, and King Edward, which was first recommended in 1931, is not only still available but was ranked at number ten in 2005 (British Potato Council figures). Further, in North America, the most popular potato cultivar, Russet Burbank, was released in 1914. It was derived by three generations of open pollination with selection from a single Chilean cultivar, Rough Purple Chili, which was introduced into the USA in 1851 (Ortiz, 2001). To compound matters, the breeding methods used by many breeding programmes have changed little in essence since Knight (1807) made the first deliberate hybridisations between different potato varieties by artificial pollination in England in 1807. Procedures remain empirical and genetically unsophisticated, relying on cycles of crossing and subsequent selection for easily observed and measured traits (*i.e.* phenotypic recurrent selection).

The rediscovery in 1900 of Mendel's published work of 1865 laid the foundation of modern genetics and crop breeding, and opened up the possibility for crop improvement based on a sound knowledge of the inheritance of economically important traits. However, compared with many other crops, progress in potato breeding appeared relatively slow and less successful during the 20th century. This has been attributed primarily to the narrow genetic base of the crop, but may also have been due its naturally high harvest index, complex inheritance patterns resulting from the possession of four sets of chromosomes, and the rather conservative, empirical approaches taken by 20th century potato breeders.

Potato fresh weight yields in Britain did, nevertheless, double from 22 to 42.5 t/ha over the forty-three year period from 1960 to 2005 (British Potato Council figures), but these increases were largely attributable to disease control, more irrigation and other improvements in agronomy, rather than new cultivars. Undoubtedly progress in improving the efficiency of potato breeding is evident in a number of areas. The application of molecular techniques as well as a fuller genetical knowledge of the important economic traits promises significant advances in the immediate future. Molecular markers are becoming an indispensable tool in plant breeding, in genetic studies and in studies of biodiversity. Increasingly, the application of molecular marker techniques will facilitate the characterisation of the potato genome, marker assisted breeding, and the characterisation of important economic traits. This brief review attempts to look forward with practical examples of improvements within modern breeding programmes.

The need for new cultivars

In Britain, the potato industry is trying to increase potato usage in an economically and environmentally sustainable way. New cultivars are required which will give economic benefits through more yield of saleable product at less cost of production, whether the potatoes are for processing or table use. They must have inherent resistances to pests and diseases. Potential new varieties also require increased water and mineral use efficiency, which give environmental benefits through reduced use of pesticides and fungicides and better use of water and fertilisers. Finally, they must help meet consumer demands for healthy, flavoursome, convenience foods.

Potato breeding has traditionally involved making crosses between pairs of parents with complementary features and this is likely to remain the main route to new cultivars. The aim is to generate a range of genetical variation on which to practise selection over a number of vegetative generations in order to select genotypes with as many desirable characteristics as possible for release as new cultivars. Increasingly, however, parents will have genes from wild species and may also be from complementary groups of germplasm to exploit hybrid vigour. For those economically important characteristics, which will still have to be handled as complex polygenic traits, SCRI has developed a potato breeding strategy which avoids the common but ineffective practice of intense early-generation visual selection between seedlings in a glasshouse and spaced plants at a seed site (Bradshaw *et al.*, 2003). Furthermore, it allows full-sib family selection to be operated on a three year cycle, an improvement on the practice of not using potential cultivars as new parents until much later when they are entered into official National List Trials. Finally, successful cultivars will also be genetically modified, provided consumers see the benefits in the use of the technology, to introduce genes not present in cultivated potatoes and their wild relatives in order to achieve further desirable improvements.

Use of wild species: marker assisted introgression

Past utilisation

Recognition of Central and South America as the centres of origin and diversity of tuber-bearing *Solanum* species, resulted in numerous collecting expeditions from those pioneered by the Russians in the 1920s to the more recent ones of the 1990s. These, in turn, led to the establishment of a number of potato germplasm collections worldwide, including the Commonwealth Potato Collection (CPC) which is held at SCRI. These collections have proved useful sources of genes for disease and pest resistance lacking in the European potato, which was derived from a few introductions of *Solanum tuberosum* from the Andes and coastal Chile from the end of the 16th century onwards. Resistance to late blight was introgressed from *S. demissum* and *S. stoloniferum*, resistance to viruses from these species together with *S. chacoense* and *S. acaule*, and resistance to potato cyst nematodes from *S. vernei* and *S. spegazzinii*. By the end of the 1980s, these wild species together with cultivated *S. tuberosum* subsp. *andigena* and *S. phureja*, were the only ones that had been used extensively in the breeding of successful cultivars in Europe (Ross, 1986) and of these, *S. demissum, S. chacoense* and *S. acaule* dominated in North American cultivars (Plaisted and Hoopes, 1989). It is, therefore, reasonable to conclude that the introgression of genes from wild species has been fairly limited and that the genetic base of modern cultivars is primarily a relatively small sample of *S. tuberosum* from the Andes and coastal Chile.

It usually took five generations to transfer a major dominant resistance gene from a wild species into a successful cultivar, and there were mixed fortunes with regard to the durability of resistance, an important issue, given the time and resources required for introgression. The major gene resistances to PVX and PVY proved durable and the *H1* gene is still effective against *Globodera rostochiensis* in Britain because Ro1 remained the main pathotype, but its widespread deployment encouraged the spread of *G. pallida*. The *S. demissum* derived R genes for resistance to late blight, in contrast, were anything but durable and breeders switched to selecting for high levels of quantitative field resistance.

Future utilisation

Further improvements in the resistance of cultivated potatoes to abiotic and biotic stresses will come from a greater use of wild species, given the wide range of habitats in which they have evolved. The screening of germplasm collections for desirable traits will continue and so will genetic studies on their inheritance, followed by introgression of desirable alleles into our European potato. These approaches will be supplemented by an increased understanding of the biodiversity present in germplasm collections which will come from molecular marker and DNA sequence data. At SCRI, a comprehensive molecular genetic analysis of the CPC is already underway to determine its overall genetic structure.

Today, virtually any potato species can be utilised for the introgression of desirable genes into *S. tuberosum*, as recently reviewed by Ortiz (1998; 2001). Furthermore, molecular marker assisted introgression offers the possibility of faster progress than can be achieved by traditional back-crossing. Genetical knowledge of the trait coupled with that of chiasmata frequency and distribution and the properties of the binomial distribution, allows the estimation of the optimal combination of population sizes and number of backcross generations (Hospital, 2003). For *S. vernei* derived PCN resistance, for example, three backcrosses with population sizes in the hundreds is realistic and could be achieved with adequate PCN screening in six years, which is much less than the five generations spread over 28 to 30 years that actually occurred at SCRI without the use of molecular markers.

Use of cultivated species from South America in base broadening

During the second half of the 20th century recognition was given to the narrow genetic base of European and North American breeding programmes. Attempts were made to widen this genetic base from the cultivated species of South America through the creation of broadly based populations of long-day adapted *S. tuberosum* subsp. *andigena* (Neotuberosum) and long-day adapted *S. phureja/S. stenotomum*. The first experiment was started by Simmonds in 1959 (Simmonds, 1969) and demonstrated that, through simple mass selection under northern latitude, long-day summer conditions, subsp. *andigena* will adapt and produce parents suitable for direct incorporation into modern potato breeding programmes. Likewise, during the period 1962-1979, Carroll (1982) employed a mass selection method to produce a population of *S. phureja/S. stenotomum* adapted to long-day North European conditions. Direct hybridisation of members of this improved diploid population with tetraploid potato cultivars via unreduced pollen grains (4*x* x 2*x* crosses) resulted in tetraploid hybrids, some of which were superior to standard tetraploid cultivars in both total and marketable yield (Carroll and De Maine, 1989).

However, to date, relatively few clones of Neotuberosum and long-day *S. phureja/S. stenotomum* have been used to any extent in the breeding of modern cultivars, despite displaying hybrid vigour in crosses with our European potato. Whilst adaptation to tubering in long-day conditions was quickly achieved, other problems remained. Neotuberosum clones lacked the regularity of tuber shape of intensively selected subsp. *tuberosum* clones, and long-day adapted *S. phureja* clones lacked tuber dormancy. Hence these populations need, and are receiving, selection for further improvements in order to achieve the original goal of direct use as parents in breeding finished cultivars. Given the biodiversity available, crosses of subsp. *tuberosum* with Neotuberosum, subsp. *andigena* and *S. phureja/S. stenotomum* will result in higher yields, better table and processing quality, improved flavour and nutritional value and some useful disease resistances.

Genetics

Since 1988, prospects for genetical knowledge based upon potato breeding have been improving dramatically and now a Dutch led international consortium plans to produce a complete physical map of the potato and to sequence the entire potato genome by the end of 2010. Two areas of immediate impact can be envisaged for this new genetical knowledge, namely in the choice of parents (*e.g.* Major gene H1 conferring resistance to *Globodera rostochiensis* Woll.) and in the selection of offspring (*e.g.* resistance to Tobacco Rattle Virus), examples of these will be presented.

Parents

The choice of parents is all important as breeding can never simply be a numbers game. Crossing the 4000 cultivars listed in the World Catalogue of Potato Varieties (Hils and Pieterse, 2005) in all possible combinations would generate 7,998,000 progenies for evaluation, and raising 500 seedlings of each would give a staggering total of 3,999,000,000, compared with the 100,000 raised in a large modern breeding programme. In contrast, a trait assessment of 4000 cultivars is feasible, and so is a genotypic assessment of diversity and content with molecular markers. Hence breeders can now think in terms of capturing allelic diversity in a smaller core set of parents and of using association genetics to choose parents genotypically as well as phenotypically. They can also use genetic distance based on molecular markers to complement co-ancestry/pedigree analysis in order both to avoid closely related parents, and hence inbreeding depression, and to ensure genetical variation for continued progress.

Multiplex parents

Research at SCRI has also sought to develop a range of parents with multiplex levels of desirable major genes. The *H1* gene is extremely effective at reducing populations of the potato cyst nematode *Globodera rostochiensis* throughout the UK and Europe. An important breeding objective is to assess the copy number of the *H1* gene in potato breeding lines, permitting clones possessing high dosages of the *H1* gene to be identified and employed as breeding parents. Triplex and quadruplex parental material give rise to 100% *H1* resistant progeny (Table 1). The SCRI variety Spey is known to be triplex for the *H1* gene and the recent SCRI variety Vales Sovereign was derived from the multiplex programme with both parents known to be duplex and is currently undergoing evaluation for *H1* gene dosage.

However, the identification of genotypes that are triplex and quadruplex for *H1* is difficult and time-consuming, requiring extensive resources during the lengthy process of phenotyping

Table 1. Proportion of genotypes with different status of H1 gene in seedling progenies.

Male parent - crossed to:	Susceptible parent	Genotypes of progeny	Phenotypes of progeny
Simplex = 1 *H1*gene	Nulliplex = 0 *H1* genes	½ Null, ½ Sim	½ Susc., ½ Resistant
Duplex = 2 *H1*genes	Nulliplex = 0 *H1* genes	⅙ Dup, ⅔ Sim, ⅙ Null	⅙ Susc., ⅚ Resistant
Triplex = 3 *H1*genes	Nulliplex = 0 *H1* genes	½ Dup, ½ Null	All resistant
Quadruplex = 4 *H1*genes	Nulliplex = 0 *H1* genes	All Dup	All resistant

derived progenies from crosses between *H1*-bearing and susceptible parents. SCRI has utilised markers from previous *H1* mapping and cloning efforts to develop quantitative single nucleotide polymorphism (SNP) markers flanking the *H1* gene that can be used indirectly to measure gene dosage. PCR product sequencing from different *H1* resistant and susceptible genotypes and comparison of sequence polymorphism data has been used to develop SNP markers whose dosage can be measured. We have employed a Pyrosequencing method to quantify relative levels of different sequence variants in a DNA sample, and applied this to a range of material with known dosages of the *H1* gene. Currently, these molecular markers distinguish between presence and absence of the *H1* gene, and moreover, can be used to measure *H1* gene dosage in some of the material assayed. Results presented demonstrate the potential of this approach for selecting for multiplex resistance status without lengthy testcross programmes and have direct application in potato breeding programmes. This method will be refined as more sequence information at the *H1* locus becomes available. The procedures used show promise for application to other major genes common within conventional potato breeding programmes *e.g.* resistance to potato viruses Y and X.

As genetical knowledge accumulates, it will be possible to choose parents for use in pair crosses with complementary gene contents. Major genes have already been mapped for flesh, skin and flower colour, for tuber shape and eye depth, and for resistances to late blight, nematodes, viruses PVX, PVY and PVA, and wart. Quantitative Trait Loci (QTLs) with genes of large effect have also been mapped for maturity and resistances to late blight, potato cyst nematodes and virus PLRV. In contrast, many economically important traits still appear to be complex polygenic traits and these include dormancy, dry matter and starch content, fry colour, resistance to *Erwinia*, tuberization and yield. For these traits, breeders will still have to rely on trait data and use offspring-midparent regressions to determine heritability and crossing strategy. It is not clear how quickly information about expressed genes and gene sequence data will translate into genes for use in breeding programmes. But again, it can not simply be a numbers game. If the potato turns out to have 30,000 genes, as few as two variants per locus (one desirable, the other undesirable) would generate a large enough number of combinations to be infinite for all practical purposes. Hence, in the genomics age, gene discovery will need to be targeted at those genetic loci likely to have the biggest social and economic impact, and understanding the genetical control of key biochemical pathways could result in the fastest progress in the immediate future.

Molecular marker assisted selection

As knowledge increases about the number and chromosomal locations of genes affecting economically important traits, breeders should be able to design better breeding programmes. As well as selecting parents that complement one another genotypically, they will be able to determine the seedling population size required for certainty of finding the desired genotype, and more realistically, the number of cycles of crossing and selection required before this is achievable in practice in the size of population they can handle. A big impact on the efficiency and rate of progress would be the identification of superior clones genotypically as seedlings in the glasshouse and the use of modern methods of rapid multiplication to progress them to commercialisation, whilst also using them as parents in the next cycle of recurrent selection. This would require molecular-marker assisted selection or preferably direct recognition of the desired gene so that the marker-trait association is never lost through genetic recombination.

The use of MAS can significantly speed up selection. Other advantages of molecular marker techniques are that they make it possible to select simultaneously for more than one characteristic in a plant. They can be used to select genotypes with a particular quality or resistance gene without exposing the plant to the pest or pathogen in question.

However, the current cost of applying these techniques is high, which means that for many breeding programmes they may currently be unaffordable but as more useful markers become available, so the costs may reduce.

Progress is expected to increase and a number of possibilities can now be found in the literature. However, to date, the only good practical example from breeding programmes that we know of is the use of SCAR markers for the PVY resistance gene Ry (from Andigena, on chromosome 11). They should be powerful markers in marker-assisted selection as they showed high accuracy for detection of the Ry gene and one marker RYSC3 was generated only in genotypes carrying Ry, namely 14 out of 103 breeding lines and cultivars with diverse genetic backgrounds (Kasai *et al.*, 2000).

TRV resistance as a target for MAS

Research at SCRI investigating the reaction to TRV infection (cause of 'spraing' or corky ring spot symptoms in tubers) of a population of potato genotypes derived from the hybridisation of a 'susceptible' x 'resistant' cultivar indicates that inheritance in this population is simply inherited. Resistance to TRV in potato cv. Record appeared to be determined by a single resistance gene. AFLP markers linked to this putative gene have been identified by bulk segregant analysis. The results of the population studies also indicate that the resistance to TRV infection is separate from the heritable system responsible for the 'spraing' response in potato tubers. The marker demonstrates the potential of MAS in the case of disease resistance that at present relies on field trials over a number of years for a virus that is often distributed in patches associated with the distribution of its nematode vectors within sites, making assessment difficult and sometimes unreliable. The AFLP band associated with TRV resistance in segregating population is arrowed in Figure 1. However, the marker is not diagnostic of resistance in a wider range of germplasm and hence further research is required.

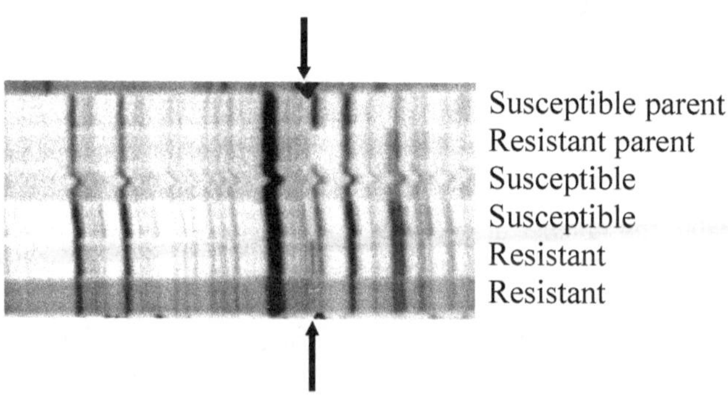

Figure 1. AFLP band associated with TRV resistance.

Conclusions

Recent research has dramatically increased the efficiency of conventional potato breeding, with emphasis now placed more on genotypic selection than on phenotypic selection as previously. With the development of molecular techniques, the next improvements in modern potato breeding programmes will be the application of molecular marker systems to important economic traits to aid selection. With the identification and localization of important genes, the cloning of genes and subsequent genetic transformation of potatoes is becoming a reality if the public perception of the technology changes to one of understanding and acceptance. The ability to meet the targets of an increasingly demanding industry and also of increasingly demanding consumers is becoming possible within the context of a sustainable agriculture system.

Acknowledgements

We thank Scottish Executive Environment and Rural Affairs Department for funding. We also wish to acknowledge the contribution of colleagues at SCRI, including G. Bryan, K. McLean and G. Ramsay.

References

Bradshaw, J.E., M.F.B. Dale and G.R. Mackay (2003). Use of mid-parent values and progeny tests to increase the efficiency of potato breeding for combined processing quality and disease and pest resistance. Theoretical and Applied Genetics 107: 36-42.

Carroll, C.P. (1982). A mass-selection method for the acclimatization and improvement of edible diploid potatoes in the United Kingdom. Journal of Agricultural Science, Cambridge 99: 631-640.

Carroll, C.P. and M.J. De Maine (1989). The agronomic value of tetraploid F1 hybrids between potatoes of group Tuberosum and group Phureja/Stenotomum. Potato Research 32: 447-456.

Hils, U., and L. Pieterse (2005). World Catalogue of Potato Varieties. Agrimedia GmbH, Bergen/Dumme, Germany.

Hospital, F. (2003). Marker-assisted breeding. In: H.J. Newbury (Eds) Plant Molecular Breeding. Blackwell, Oxford: 30-59.

Kasai, K., Y. Morikawa, V.A. Sorri, J.P.T. Valkonen, C. Gebhardt and K.N. Watanabe (2000). Development of SCAR markers to the PVY resistance gene Ryadg based on a common feature of plant disease resistance genes. Genome 43: 1-8.

Knight, T.A. (1807). On raising of new and early varieties of the potato (Solanum tuberosum). Transactions of the Horticultural Society of London 1: 57-59.

Ortiz, R. (1998). Potato breeding via ploidy manipulations. In: J. Janick (Ed.) Plant Breeding Reviews. John Wiley & Sons, New York: 15-86.

Ortiz, R. (2001). The state of the use of potato genetic diversity. In: H.D. Cooper, C. Spillane and T. Hodgkin (Eds.) Broadening the Genetic Base of Crop Production. CABI Publishing, Wallingford: 181-200.

Plaisted, R.L. and R.W. Hoopes (1989). The past record and future prospects for the use of exotic potato germplasm. American Potato Journal 66: 603-627.

Ross, H. (1986). Potato Breeding - Problems and Perspectives. Advances in Plant Breeding 13, Paul Parey, Berlin and Hamburg.

Simmonds, N.W. (1969). Prospects of potato improvement. Scottish Plant Breeding Station Forty-Eighth Annual Report 1968-69, pp. 18-38.

Curriculum vitae – Dr M.F.B. Dale

Finlay Dale has an honours degree in Agricultural Science from Edinburgh University and a PhD in Plant Biology from the University of Birmingham. His doctorate degree, based at Birmingham University and the Plant Breeding Institute in Cambridge, examined wild and primitive species of sugar beet, their relationships and possible evolutionary directions and factors affecting microspeciation within the genus *Beta*. He joined the Scottish Plant Breeding Institute in 1979 as a research scientist and subsequently transferred to the Scottish Crop Research Institute in Dundee in 1989, researching aspects of breeding potatoes and important disease/quality traits of the potato. Research areas include investigating resistance to potato cyst nematode, the inheritance and effectiveness of new sources of resistance to *Globodera pallida* in wild *Solanum* species from the Commonwealth Potato Collection gene bank based at SCRI. The research evaluates the resistance and applies molecular markers to facilitate gene introgression into modern cultivars. Recent research has investigated the epidemiology of Tobacco Rattle Virus, an important virus in potatoes, particularly processing potatoes. The research has important implications for the seed potato industry in the UK, Europe, and further afield. Research is currently phenotyping potato populations to develop molecular markers for newly identified major gene resistance to facilitate selection and possible GM approaches. Finlay Dale is one of the lead potato breeders at SCRI with joint responsibility for a number of research and commercial breeding programmes. He is a member of the Association of Applied Biologists and the Plant Physiology and Crop Improvement committee and is secretary to the potato section of the Scottish Society for Crop Research. He is also an editor for the scientific journal Annals of Applied Biologists from 1998 and was honorary lecturer at Dundee University 1994-1999. He is author of over 50 scientific publications.

Company profile – Scottish Crop Research Institute

The Scottish Crop Research Institute (SCRI) is the lead centre in the UK for research on potatoes, barley and soft fruit crops. It also carries out research on a wide range of temperate, sub-tropical and tropical crops, *e.g.* cassava, groundnut, mahogany and coffee. A comprehensive range of skills are available, from fundamental studies on genetics and physiology, through agronomy, biochemistry and pathology, to glasshouse and field trials. A broad multidisciplinary approach to fundamental and strategic research, and technology transfer, are strengths of the SCRI. Programmes span the disciplines of genetics and breeding, molecular and cellular biology, biotechnology, plant pathology (bacteriology, entomology, mycology, nematology and virology), plant physiology and cell biology, environmental science, plant chemistry and biochemistry, agronomy, molecular ecology, vegetation dynamics, bioremediation, serology, physics, mathematics, bioinformatics and statistics. Genetics and enhanced breeding of selected crops, and biotechnology, lie at the core of our research, development and training programmes. Breeding programmes at SCRI, in collaboration with many commercial partners, are responsible for new crop varieties that are widely released throughout the World. Research programmes are:

1. **Environment Plant Interactions:** processes in the soil to the plant environment, providing an integrated understanding of managed agro-ecosystems.
2. **Genetics:** using modern approaches in breeding and genetics this programme delivers improved crop plants that fulfill a range of end-user requirements. Research includes Genome

biology, Genetics and breeding, Genes and development, Biodiversity, Bioinformatics and statistics.

3. **Plant Pathology:** research in molecular and cellular areas to that of pest and pathogen populations. Objectives are to understand disease processes and provide knowledge and strategies for crop improvement through durable resistance and sustainable disease control strategies.
4. **Quality, Health and Nutrition**: research investigating the processes that regulate specific quality and nutritional parameters of our important commodity crops.

Stacking of resistance genes in potato by cisgenesis instead of introgression breeding

E. Jacobsen and R. Hutten
Wageningen University and Research Centre, Laboratory of Plant Breeding, P.O. Box 16, 6700 AA Wageningen, The Netherlands

Abstract

Potato breeding is an important activity to improve potato production for different applications. There are two ways of potato breeding: 1. selection of new varieties which are based on a cross breeding program and 2. improvement of existing varieties. Until recently, the improvement of existing varieties could only be obtained by classical mutation breeding. Nowadays, existing varieties can also be improved by genetic modification with transgenes. Examples are, the introduction of virus resistance by coat protein genes or insect resistance by Bt genes. This way of breeding by transgenes has not been accepted in the potato market and is, therefore, not used by potato breeders. The opposition of the non-governmental organizations (NGOs) and the complicated regulation for genetic modification (GM-regulation) in the different countries are the main bottle neck. The GM-regulation in the past has been based on risk assessments using transgenes coming from other organisms like bacteria or viruses which are not belonging to the natural gene pool of the plant breeder. Sequence information combined with the development of efficient gene isolation techniques enable breeders to use plant own genes and genes of crossable species to produce new varieties. We call these natural plant genes cisgenes to distinguish them from transgenes. These cisgenes belong to the same gene pool as those which are available to the classical plant breeder. However, because new gene transfer techniques are used, this way of breeding has been placed under the EU-legislation for GMOs which hampers the use of this advantageous approach. For resistance breeding, this very efficient cisgenic approach could replace and/or complement introgression breeding which is normally used for stacking of resistance genes in case single resistances are easily overcome by the pathogen as in the case of, for example, *Phytophthora infestans*. Introgression breeding has a number of disadvantages in comparison with cisgenesis. The most important ones are: 1. multiple (back) cross generations are needed to come to domestication and stacking of resistance genes by introgression breeding, 2. this process is always accompanied by linkage drag with hundreds of other genes of the wild species, and 3. existing varieties cannot be improved by introgression breeding. If the genes of interest are cloned, the cisgenic (or intragenic) approach is for complex crops like potato, banana and cassava an alternative for introgression breeding. It enables 1. single step domestication and stacking of resistance genes without linkage drag, and 2. improvement of existing varieties. Because of the use of genes from the breeders gene pool and the parallel with introgression breeding, exemption of the Directive 2001/18/EC for the cisgenic approach is strongly recommended if marker-free cisgenes are used.

Keywords: introgression breeding, cisgenesis, transgenesis, improvement of existing varieties, exemption of EU-directive 2001/18/EC

Introduction

The potato breeding process is getting more and more complex in order to create agriculturally and/or qualitatively improved new varieties. At the onset of potato breeding, the potato breeder needed not more than hundred seedlings to select an improved potato variety which was better adapted to our day length and other environmental circumstances. The occurrence of late blight more than 150 years ago, stimulated potato breeding involving crosses with wild species. Over the years, the breeding approach using wild species, nowadays mentioned introgression breeding, has been worked out in more detail (Jacobsen and Ramanna, 1999). It was successful in respect to resistance to among others potato viruses, potato cyst nematodes and wart disease (Ross, 1986). In case of *Phytophthora infestans* major resistance genes originating from *Solanum demissum, S. edinense* and *S. stoloniferum* were broken down rapidly and resistance breeding by R-genes was discarded for a long time. However, the alternative approach using field resistance, which is based on polygenes causing partial resistance, was not successful at all. Despite the lack of success, this quantitative approach is still advocated by many researchers worldwide.

The number of different applications of potato in different products has extended tremendously during the last 50 years. Therefore, besides agricultural improvements, improvement of quality traits became a necessity. This means that at present, because of the multiple traits desired, new varieties are found much less frequently (1 variety on 30,000 - 100,000 seedlings) and in addition pre-breeding, for crossing parents, is an important activity (Ross, 1986).

Nowadays, the new developments in genomics are going fast enabling genetic mapping, including localization of monogenic traits as well as polygenic, quantitative, traits by QTL's. This genetic mapping is an important step to come to more effective marker assisted selection and breeding and is stimulating indirect selection to speed up the classical breeding process.

For 15 years, transformation of potato, using cloned genes is possible (Jacobsen, 1999). This type of breeding is comparable with mutation breeding to improve existing varieties with only one or a few new traits. The most important way of transformation is by using *Agrobacterium tumefaciens*. At the early phase of GM- research and GM-breeding only transgenes, originating from other organisms (for example: bacteria and viruses)than crossable plants, were available for variety improvement.(Jacobsen, 1999). For selection of transformed cells and regeneration of plants containing the inserted gene of interest, bacterial selection markers, coding for antibiotic resistance or herbicide resistance, were a specific need.

Objections in society prevent the cultivation of potato, GM-varieties in Europe and world wide. The NGO's with their strong lobby were able to influence the market in such a way that "no GM certificates" were introduced in order to assure for the processing industry as well as the retail the availability of GM-free fresh potatoes and processed products. This successful NGO activity has stopped the breeding and introduction of GM potato worldwide. The main objections against GM crops concerned the use of selection markers like antibiotic and herbicide resistance genes, questions around human health risks related to toxicity and allergy, and ethical aspects (Jochemsen, 2000). In general, the introduction of transgenes, originating partly or totally from another organism or non crossable species, concerns genes from other gene pools than the classical breeders or crop growers in the agro-ecosystem are familiar with (Anonymus, 2001). It is therefore reasonable that the introduction of such transgenes for domestication in a crop has to be followed more carefully than the use of genes belonging to the normal gene pool of the breeder and the grower which do have a long tradition of safe use.

In recent years cloning strategies for all kind of useful plant genes have been developed not only for model plants, like *Arabidopsis*, but also for important crops like maize, rice, Brassica rapa, tomato and potato (Huang *et al.*, 2005). Complete genome sequencing projects have been started or even finished. This means that for most important crops many or all genes can be cloned and will be available with their different alleles for research and breeding application. Using a GM approach, we refer to these isolated genes coming from the crop plant itself or from crossable wild as cisgenes and not as transgenes because they arise from the same gene pool and resources that are used by the classical breeder (Schaart, 2004; Schouten *et al.*, 2006a and 2006b).

Different from introgression breeding is that cisgenes, introduced by genetic modification, despite their origin are subject to the Directive 2001/18/EC for GM crops (Anonymus, 2001). This discrepancy is the subject of this paper. Cisgenesis is connected with solving problems in resistance breeding where stacking of resistance genes is important, especially in cases where single gene resistances can rapidly be broken down by the pathogen. Comparison between transgenesis, cisgenesis and introgression breeding brings the insight that, with respect to the outcome, cisgenesis resembles much more introgression breeding than transgenesis because of the gene pool source (Schouten *et al.*, 2006a and 2006b).

Introgression breeding

Introgression breeding is a very important breeding activity. The use of wild species has been a significant improvement for plant breeding in terms of broadening the genetic variation for disease resistances and quality traits. The detection of the centers of origin of crop plants, the many expeditions to collect potato germplasm in the Andean region as center of origin, the foundation of gene banks for the most important crops, the evaluation of collected plant material for important traits, and research related with interspecific hybridization are a few major aspects directly related with introgression breeding. The increase of biodiversity, including crossable wild species, was highly needed in order to feed plant breeding with new natural genetic variation. The use of genetic variation found in gene banks, including interspecific crosses followed by backcrosses, is nowadays still the heart of modern potato breeding (Ross, 1986).

Biotechnology has broadened this way of breeding by adding 1. *in vitro* techniques like embryo rescue, regeneration, propagation and inter/intraspecific protoplast fusions; 2. genomics by the development of genetic maps with molecular markers, enabling marker assisted breeding (Jacobsen, 1999); and 3. genetic modification of plants by using transgenes obtained from non-crossable species and even other organisms. The GM approach with transgenes was new and a separate EU-legislation has been developed, focused on risk prevention. It is of importance to realize that in addition to modification with transgenes, the technique of protoplast fusion between crossable species was included in the GM definition (Anonymus, 2001), but that the breeding products were exempted from the GM-legislation. The same was the case with natural and induced mutations.

One of the drawbacks of introgression breeding is the long time of pre-breeding needed to make an optimal use of the genetic resources. Introgression breeding is a process of interspecific hybridization followed by backcrosses with the cultivated species under the simultaneous selection for the trait of interest. The usefulness of the trait of interest is highly dependent on the simultaneous linkage of traits with negative impact. This phenomenon is called "linkage drag" which can be removed by crossing-over during meiosis (Jacobsen and Ramanna, 1999). The removal of such negative traits is dependent on the distance to the trait of interest and on the recombination frequency during meiosis. Often, it is observed that around the location of introduced traits from wild species after introgression breeding, crossing-over cold spots are found, preventing the removal of negative traits. There are many examples mentioned in literature showing that desired traits from related species did not came available for variety breeding because of linkage with late maturity, small fruits or other problems (Jacobsen and Ramanna, 1999). In other cases, where introgression breeding is successful, the end result is still introgression of a trait surrounded by many alleles from other genes from the donor species. In case of stacking resistance genes from different sources by introgression breeding linkage drag is representing even a greater barrier. Linkage drag is not always only connected with negative traits. It is also a source of new alleles which can be of importance for heterosis effects on yield.

Mutation breeding

The results of transgenesis and cisgenesis are comparable with the outcome of mutation breeding, *i.e.* adding or removing a specific agricultural trait. A mutation in plants is defined as a sudden genetic change of very different kind in the idiotypic constitution of sporophytic and gametophytic plant tissues (Harten, 1998). Mutations are the ultimate source of genetic variation, and together with selection and recombination of genes, the most important factors in evolution. Mutations of plant traits are by definition single cell events and, therefore, in vegetatively propagated crops always accompanied with the problem of chimerism. In a heterozygous polyploid crop like potato not many induced mutations have been important for breeding. However, phenotypic changes like tuber color, maturity type and tuber shape have been used in variety breeding as well as biochemical changes like starch composition in the *amf*-mutation for amylose-free starch (Harten, 1998). There are various types of mutations possible *i.e.* 1. intra-genic mutations within individual genes; 2. inter-genic or structural mutations within chromosomes; and 3. genome mutations with changes in chromosome number (polyploidy, haploidy and aneuploidy). At the DNA level, three types of mutations are discerned, 1. point mutations or base substitutions; 2. frameshift mutations based on single nucleotide insertions or deletions; and 3. macro-mutations including rearrangements of chromosome segments such as inversions, deletions, duplications, translocations or insertions by T-DNA in GM plants or by natural transposons as described in maize and petunia (Harten, 1998). Mutations can occur spontaneously or they can be induced by chemical treatments or radiation.

Breeding with GM plants for the introduction of specific traits is highly comparable with mutation breeding for dominantly inherited mutations, but the use of gene silencing is comparable with recessively inherited loss of function mutations. A big difference between mutation breeding and the use of transgenes is the gene source. Transgenes are introducing sudden changes but they are originating from genes belonging to other organisms or non-crossable species. The use of cisgenesis is introducing sudden changes with the insertion of plant own genes which are originating from the same plant species or from crossable species. In this sense, cisgenesis is a parallel with macro-mutations, including very small chromosome segments. In this case, it is restricted to one gene and is comparable with a translocation without linkage drag in classical breeding.

Transgenes and EU-legislation

Transgenesis is the transfer of transgenes into crop plants to introduce new traits for improvement of existing varieties or to create improved crossing parents for breeding new varieties. Transgenes, originating from other organisms like bacteria and viruses, were at the beginning of the GM breeding era the main gene sources available. Therefore, the Directive 2001/18/EC (Anonymus, 2001) for the introduction of GM crops into the market has mainly been focused on and inspired by transgenes. These transgenes were originating from a new gene pool from other organisms. The use of bacterial antibiotic resistance or herbicide resistance genes was needed for selection of individual transformation events at cellular level and after plant regeneration at shoot level. The assessment developed for GM crops in all stages of variety development is only focused on prevention of risks. The release of GM crops in Europe is totally dependent on risk assessments and not on risk-benefit assessments by making a comparison between the old and the improved new situation. In fact, we are assessing domestication of transgenes into the crop plant and its agro-ecosystem. The introduced transgenes at the beginning of the GM breeding approach partly or totally originate from other organisms and are more comparable with horizontal gene transfer (Schmidt and Hankeln, 1995).

Cisgenesis and stacking of resistance genes

Cisgenesis is an important new possibility in GM plant breeding. The main reasons for this development are: 1. the new possibilities of transforming crop plants without (bacterial) selection marker genes in seed propagated crops by mixed inoculation and even in vegetatively propagated crops like potato by marker-free vectors (Vetten *et al.*, 2004); and 2. cloning of important genes in crop plants. In seed propagated crops, mixed inoculation by using simultaneously two *Agrobacterium* strains, one containing the selection marker and the other one the gene of interest, is resulting in single cells with independent insertion events of both TDNAs with simultaneous expression of both the selection gene and the agricultural gene. After plant regeneration, one sexual step is needed to separate both insertions via meiotic segregation, resulting in different classes of offspring plants such as the class of GM plants expressing the agricultural trait but without selection marker. In vegetatively propagated crops, which like potato are frequently highly heterozygous, a sexual step is not possible if an existing variety has to be improved by addition or silencing of certain traits. For this situation, a marker-free vector system has been developed (Vetten *et al.*, 2004). Cisgenesis is made possible in vegetative propagated crops if such a marker-free vector system is combined with the use of only cisgenes, *i.e.* natural genes from the crop plant itself or from crossable related species coding for agricultural traits (Jacobsen *et al.*, 2005). In practice, cisgenesis will certainly be used in resistance breeding of vegetatively propagated complex crops like potato, banana and cassava. In many cases resistance genes have to be found in related wild species and introgression breeding is needed to introduce them into cultivated plants. As described in the introgression breeding section, the end result is multiple step domestication of resistance genes including linkage drag. This approach generates multiple problems if stacking of genes is needed. Another observation in resistance breeding is that resistance genes are found in clusters on the different chromosomes. It means that, if more resistance genes from the same cluster are needed in one variety, a combination of these genes from different sources has to be realized by crossing-over events between different plants at the same chromosomal location. In potato a few more resistance genes can be combined without the need of such a crossing-over event because of its polyploid nature, but accumulation of linkage drag problems at the same chromosomal location from different species is the price that has to be paid.

The alternative cisgenesis approach, being single step domestication of one or more cloned genes without linkage drag, is highly attractive. Cisgenesis has the possibility to stack agricultural genes from different sources even if they are located at the same position in the genome without the problems of linkage drag and/or crossing-over dependence.

Insertion of cisgenes is random and comparable with phenomena like random insertion in already released transgenic GM plants, natural transposons and translocations in classical breeding as described in the mutation section. These random insertions are very important, it enables us to select for insertion places with the best expression of the introduced gene but simultaneously without additional disturbing or undesired side effects. Transformants with major negative effects and insufficient expression are already removed at the beginning of the selection process in the growth chamber and the glasshouse. The intensive classical selection process in the field with cisgenic GM plants at multiple sites, resulting in an improved variety, is protecting us from small negative side effects, including those of biochemical changes like higher glycoalkaloid content in potato. Estimation of the glycoalkaloid content is already a routine in the classical breeding process of potato because of the frequent use of wild species which are containing higher amounts of this class of toxins.

Cisgenesis is, because of its gene source, highly comparable with classical introgression and mutation breeding.

Conclusion

The classical approaches of resistance breeding, despite their success have major disadvantages connected with multiple step domestication of resistance genes from crossable species, linkage drag, and with stacking genes. The use of molecular markers is speeding up these processes by indirect selection but is not solving linkage drag problems. For combining resistance genes at the same chromosomal cluster, crossing-over is the major bottle neck. In vegetatively propagated crops like potato, introgression breeding cannot be connected with the direct improvement of existing varieties.

The new development with cisgenes is a powerful alternative:

1. Despite of being a GM approach it is expected that, in addition to protoplast fusion and mutation, cisgenes will be exempted from the transgenes based Directive 2001/18/EC (Anonymus, 2001);
2. Cisgenesis is a single step domestication approach without linkage drag and insertion related negative side effects can be overcome when enough transformation events are present for selection;
3. Stacking of resistance genes as resistance strategy can more easily be made by inserting cisgenes which is complementing or replacing introgression breeding;
4. Existing varieties can be adjusted in one step by the introduction of cisgenes.

References

Anonymus, (2001). Directive 2001/18EC of the European Parlement and of the Council of 12 March 2001 on the deliberate release into the environment of genetically modified organisms and repealing Council Directive 90/220/EEC. Off J Eur. Comm 106:1-38.

Harten van, A.M (1998). Mutation Breeding, Theory and Practical Application. Pp 353. Cambridge University Press, Cambridge, UK

Huang, S,. E.A. Vossen van der, H. Kuang, V.G. A.A. Vleeshouwers, N. Zhang, T.J. Borm, H.J. Eck van, B. Baker, E. Jacobsen and R.G.F. Visser (2005). Comparative genomics enabled the isolation of the R3a late blight resistance gene in potato. Plant J. 42:251-261.

Jacobsen, E and M.S. Ramanna (1999). Introduction to plant breeding II: Genetic variation. Part I: Genetic modification *in vivo*. Lecture notes. Wageningen Agricultural University, pp 129.

Jacobsen, E (1999), Introduction to plant breeding II: Genetic variation Part II: Genetic modification *in vitro*. Lecture notes, Wageningen Agricultural University, pp. 135.

Jacobsen, E., T.H. Park, S. Vossen van der, V.G.A.A. Vleeshouwers and R.G.F. Visser (2005). Towards Phytophthora resistant GMO-potato varieties exempted from GMO-regulations. In: A. Ritter ans A. Carrascal (Eds), Abstracts of papers and posters of the 16th Triennual Conference of the EAPR European Association of Potato Research July 17-22, Bilbao, Spain: 243-245.

Jochemsen, H (2000). Ethical judgement of genetic modification. In: H Jochemsen (Ed). Test and confine. An ethical and political judgement of the modern biotechnology (in Dutch). pp 88-95. Christen Unie, Nunspeet, the Netherlands.

Schaart,J.G (2004). Towards consumer-friendly cisgenic strawberries which are less susceptible to Botrytis cinerea. PhD-thesis Wageningen University, ISBN:90-8504-104-x, Wageningen, The Neherlands, pp 117.

Schouten,H.J.,F.A. Krens and E. Jacobsen (2006a). Cisgenic plants are similar to traditionally bred plants. EMBO Reports. In press

Schouten, H.J., F.A. Krens and E. Jacobsen (2006b). Cisgenic plants should be excluded from GMO regulations. A Comment. Nat. Biotechn. In press.

Ross. H (1986). Potato breeding- Problems and perspectives, Berlin, Hamburg: Verlag Paul Parey.

Schmidt, E.R and Th. Hankeln (1995). Transgenic organisms and biosafety: Horizontal gene transfer, stability of DNA, and expression of transgenes. Springer pp 335.

Vetten, N., M.A. Wolters, K. Raemakers, I. Meer van der, R. Stege ter, P. Heeres and R.G.F Visser, (2003). A transformation method for obtaining marker-free plants of a cross-pollinating and vegetatively propagated crop. Nat. Biotechn. 21:439442

Curriculum vitae – Prof. Dr. Ir E. Jacobsen

Prof. Dr. Ir. E. Jacobsen was born on 20.05.1947 in Arnhem, the Netherlands. He finished his BSc and MSc study in 1974 on Plant Breeding at Wageningen University, the Netherlands and his PhD study on Modern Potato Breeding Techniques in 1978 at the Friedrich Wilhelm University in Bonn, Germany.

His first job from 197-1978 was on modern potato breeding research at the Max Planck Institute for Plant Breeding in Cologne, Germany. In the period 1978-1988, he was staff member at the department of Genetics of the Groningen University, the Netherlands. Genetic dissection in pea of the nodulation of nitrogen fixing root nodules by selection of induced mutants and development of biotechnological techniques for unraveling starch biosynthesis in potato genetics were some main activities. The isolation of the amylose-free potato mutant was a major success. In 1988, he was appointed at Wageningen University as professor in Plant Breeding. GMO research on starch biosynthesis was a key issue. Two amylose-free GMO potato varieties were produced out of the research in the nineties. At a certain moment over 1500 ha of these GMO potatoes were grown in the Netherlands before the NGOs were successful with their protests. In 1991, he started the successful graduate school Experimental Plant Sciences and was director of this school until the end of 1998. In the period 1997-2001, he was director of the Department of Plant Sciences of Wageningen University and in the period 2001-2004, he was science director of the Plant Sciences Group of Wageningen University and Research Center. At the moment, he is professor of Plant Breeding and chief scientific officer of the Plant Sciences Group. The coming years, he will work on the acceptance and introduction of cisgenesis and cisgenic crop varieties, including potato. The cisgenic crops have, in parallel with mutants and protoplast fusions between crossable species, to be exempted from the Directive 2001/18/EC on GMOs.

Curriculum vitae – Dr. Ir. R.C.B. Hutten

Dr. Ir. R.C.B. Hutten was born on 04.11.1960 in Goirle, The Netherlands. He finished his MSc study in 1988 on Plant Breeding at Wageningen University, The Netherlands. In 1994 he defended his PhD thesis on "Basic aspects of potato breeding via the diploid level" at Wageningen University. From 1987 onwards he is involved in applied potato breeding research at the Laboratory of Plant Breeding, Wageningen University. The most important topics of his research are related to pre-breeding, late blight, potato cyst nematodes and rootknot nematodes. He is a member of the management team of the Dutch Phytophthora Umbrella Programme.

Profile – Wageningen University and Research Centre: Plant Sciences Group

The Plant Sciences Group is part of Wageningen University and Research Centre, a cooperation between Wageningen University, research institutes, a training and advisory centre and the Van Hall Larenstein Polytechnic. The University and research institutes work together closely in five areas of expertise: Plant Sciences Group, Animal Sciences Group, Social Sciences Group, Environmental Sciences Group and the Agrotechnology & Food Sciences Group.

The Plant Sciences Group is a collaboration between: Wageningen University Plant Sciences, Plant Research International BV (PRI) and Applied Plant Research BV (PPO).

Within Wageningen UR, the Plant Sciences Group brings together science education and scientific, strategic and applied research in the fields of biology of plants in relation to their environment, of plant-related organisms, and of plant production. The Plant Sciences Group provides expertise and products in the field of healty foods, sustainable green production chains, biodiversity and plant and environment to students, governmental institutions and industry.

Sustainable and organic production

Potatoes and the environment: An overview

Melvyn F. Askew
Head of Agriculture and Rural Strategy Group, Central Science Laboratory, York, UK, YO41 1LZ, Visiting Professor Harper Adams University College, United Kingdom

Abstract

Whilst the potato crop ranks fourth in the world in terms of tonnage of production, the ground area occupied by the crop is relatively small. Intensive production in potato growing areas allied with a market demand for clean-skinned, disease free potatoes of specific size/grade, focuses the risks to the environment from potato production to a significant extent. Whilst some risks can be minimised through careful agronomic approaches, others, including cultivations, crop protection and harvesting remain as environmental challenges.

Keywords: environment, agronomy, soils

Introduction

The European or White Potato (*Solanum tuberosum*) was introduced to Europe approximately 500 years ago. It is now grown as a mainstream crop in all European countries (see Table 1) and virtually in all other countries in the world.

The market for potatoes and potato products has differentiated to a very significant extent within each of the three basic potato market sectors, fresh potatoes; processed potatoes and potato seed tubers. Usage of potatoes as a primary feedstock for the non-food sector occurs in starch production, especially in France, Germany, Denmark and in the Netherlands. Additionally, biodegradable and fully compostable biopolymers, *e.g.* "Solanyl" (www.biopolymers.nl), are produced from potatoes.

Additionally fresh potatoes are available year around in Europe and processors of potatoes require continuous supplies to maintain factory operations.

Data, calculated from FAO Statistics, suggests that in the former EU-15, the area of arable land devoted to potato production varied from approximately 20% in the Netherlands to less than 1% in countries like Italy, Spain, Sweden or Austria. Most countries had between 2% and 3% of their arable land devoted to potatoes.

Clearly, with the wide-ranging scenarios outlined, the potential for impact of potato production and utilisation on the environment is high, although not necessarily always negative.

Potatoes and the soil

In Europe, the potato is produced as a spring planted crop. A limited number of situations arise when so-called 'early potatoes' are planted at the end of a year for harvesting by May of the following year. Nonetheless, the preponderance of planting takes place during the months

Table 1. Potato Production in the EU-25 in 2005.

Country	Area of production (1000ha)	Harvested production (1000t)	Yields (100kg/ha)
European Union	2.031,393	59.530,104	293,051
Austria	22,173	707,911	319,267
Belgium	64,325	2.653,949	412,584
Cyprus	6,500	116,000	178,462
Czech Republic	36,071	1.000,000	277,231
Denmark	40,000	1.600,000	400,000
Estonia	15,800	224,800	142,278
Finland	28,800	739,000	256,597
France	147,037	6.347,499	431,694
Germany	276,300	11.157,500	402,945
Greece	45,000	840,000	186,667
Hungary	24,826	607,220	244,590
Ireland	12,000	500,000	416,667
Italy	72,156	1.810,086	250,819
Latvia	45,900	710,000	154,684
Lithuania	63,200	1.000,000	158,228
Luxembourg	0,608	19,329	319,110
Malta	1,100	25,000	227,273
Netherlands	161,000	6.835,985	424,595
Poland	594,308	11.009,390	185,247
Portugal	80,000	1.250,000	156,250
Slovakia	24,226	381,891	157,637
Slovenia	6,260	153,131	244,888
Spain	93,,300	2.591,700	277,781
Sweden	30,540	950,000	311,057
United Kingdom	140,000	6.300,000	450,000

Source: FAOSTAT 2006

of February to May, depending upon locality, season and, to a degree, intended market for the crop.

Clearly, a bare soil surface over the winter period enhances the risks of leaching of nitrates in particular, although the extent to which this will occur is dependent upon soil type, weather conditions and previous cropping. Light, well drained soils which have been recently cropped with plants which leave significant nitrogen residues are at most risk and more so in a winter where temperatures do not fall below 0 - 4°C for any lengthy period of time.

Coenegrachts (Personal Communication) considered the operation in the seedbed preparation phase for potatoes, which created medium to high damage to soil structure to be:

- Deep cultivations after subsoiling
- Ploughing
- Clod and/or stone separation
- Bed/ridge forming

In addition, operations carried out in the growing crop especially where narrow wheels/tyres were used, created compaction and conditions for subsequent water erosion. Harvesting of potatoes has the potential to cause severe soil structural damage through compaction, particularly in adverse weather conditions.

Compaction has adverse environmental and economic effects:
- It reduces water permeability and raises the potential for water erosion; irrigation enhances this potential.
- It reduces root penetration and therefore limits plant growth and yield. It also limits the ability of roots to absorb nutrients, this in turn enhances loss of nutrients below the compacted horizon into the local eco-system.
- Compaction caused during the potato crop phase of the rotation can impact adversely on soil and crops later in the rotation.

Reduction of adverse effects associated with preparation of soils for potato planting can be achieved by careful selection of tyres or equipment; adjustment of tyre pressures; minimising the number of operations, and avoiding where possible, undertaking operations in adverse conditions.

Whilst not applicable to many "potato soils", the risk of wind erosion after seedbed preparation occur on sands, loamy sands and some peat soils. Destoning tends to enhance this risk. Experience in England and Wales suggests that, those crops which were most likely to be associated with wind erosion were potatoes and sugar beet.

In addition to environmental risks, which are directly attributable to potato cropping, issues of aspect and topography have generic effects on all crops where surface run-off of water has potential. Clearly these are generally outwith the control of the individual farmer.

Agrochemicals, potatoes and the environment

The production of potatoes in the field involves fertilisers, crop protection materials and, according to the crop and its age and intended market, haulm destruction procedures or materials. Storage of potatoes, especially longer-term storage for processing may involve the use of sprout inhibiting chemicals.

Fertilisers

Potatoes require relatively large amounts of nutrients to achieve high yield, particularly potassium. Average amounts of nutrients removed by a 50t/ha crop of potatoes are 215 kg/ha N, 83 kg/ha P and 443 kg/ha K.

Fertiliser use varies between countries according to soil type, expected yield, intensity of production, method of calculating recommendations and traditional farm practice. Average levels of nitrogen use in some countries appear to be higher than necessary.

The potential for nitrate leaching from potatoes is greater than for many other crops. Generally, information is available to farmers which aims to limit the potential for leaching, through soil analysis, recommendations relating to soil type and nutrient status and modelling to predict nutrient requirements. However, the level of information is variable across Europe. For example, in the UK, nitrogen recommendations for potatoes are based on soil mineral nitrogen index, length of growing season, and variety, but also recommend specialist guidance for specific crops. In Germany, nitrogen recommendations for potatoes are based on soil mineral nitrogen content or, if no soil analysis is done by the farmer, there is a regional information system, which gives the information for different soil types for each region. In Spain however, soil analysis seems to be less common, official fertilizer recommendations tend to be identical for all farms of a region. Split fertiliser applications optimise nitrogen use efficiency by matching nitrogen supply to the crop requirement, and can help to reduce nitrate leaching.

Pesticides including storage chemicals

Potatoes suffer from a large number of diseases, but by far the most important is Late Blight caused by *Phytophthora infestans*. This can seriously reduce yield, especially in wet years, when the tubers may also be infected. Prophylactic fungicide spraying at 10-14 day intervals is generally carried out to control the disease, with haulm destruction before harvest to prevent tuber infection. Weather-based blight warning systems exist in several countries to assist growers with timing of preventative measures. New blight-resistant SARPO varieties from Hungary may help to reduce the need for fungicide use in the future.

Other important diseases are Common Scab (*Streptomyces scabies*), Gangrene (*Phoma exigua*), Dry Rot (*Fusarium caeruleum*), and Silver Scurf (*Helminthosporium solani*).

Aphids are important pests because they spread viruses which can severely affect yield. In mainland Europe, Colorado beetle is a major pest, with potential to cause complete crop loss. In the UK, slugs are important, and in countries where potatoes are grown frequently, potato cyst nematodes are a major problem. The use of insecticides, molluscicides and nematicides to control these pests has considerable environmental implications.

Countries differ in the extent of pesticide use, but in all countries it is high compared with other crops.

Potato production systems are generally intensive with very high pesticide inputs, particularly of fungicides and insecticides, relative to other major crops. Particularly characteristic of potato crops is the widespread use of nematicides, desiccants and (in store), sprout suppressants, which are not used on other major crops such as cereals. However, the quality and type of information available on pesticide use is very variable, making comparisons between countries difficult.

Pesticide use was particularly high in The Netherlands, estimated at 18-20 kg/ha/year in the early 1990s, though there have been some reductions over the last decade, particularly in the use of "soil disinfectants". Potatoes accounted for 42% of pesticide use by weight in The Netherlands. In the UK, 1516% of the crop area was sprayed, compared with 851% for cereals. In Germany, the application index (number of applications in relation to recommended rate and area) was 8.56 for potatoes, compared with *e.g.* 3.74 for wheat.

Levels of fungicide use are very high and have tended to increase, because of the need to control Late Blight and the lack of alternative control methods. Most widely grown varieties are sensitive to this disease.

Insecticides are used largely to control aphids and Colorado beetle, and nematicides are widely used against potato cyst nematodes (PCN) in northern Europe. Of the two prevalent species of PCN, some potato varieties are almost completely resistant to *Globodera rostochiensis;* however, only partial resistance to *Globodera pallida* exists. *G. pallida* is more difficult to control than *G. rostochiensis* because it has greater genetic diversity and a more prolonged period of hatch, and because virulent populations have been selected through the inappropriate use of partially resistant cultivars.

Weed control is less important than in some other crops, as, once the crop grows rapidly it produces a dense canopy. However, herbicides are used in the early stages to prevent weed competition and to ease harvesting. The potato crop also causes major problems in following crops by the production of volunteers. With the increased frequency of wet conditions during harvesting and reduced frequency of frost periods during winter this problem has increased, causing more herbicide use in the following crops.

Chemical haulm desiccants are applied to the majority of potato crops across Europe. The most widely used desiccants are sulphuric acid and diquat. Mechanical haulm destruction can also be used either alone, or a combination with desiccants.

In case of storage at elevated temperatures, *e.g.* for processing, a number of active ingredients have been used as sprout suppressants, including tecnazene; maleic hydrazide, CIPC (chlorpropham), and carvone (from caraway seed). However, CIPC is now the only significant sprout suppressant. However, it is particularly persistence and can have adverse impacts on many stored crop seeds. As an alternative approach to sprout control, new genotypes of potato which do not accumulate reducing sugars after periods of cold (c 4°C) storage are being developed especially for processing purposes.

Data on the ecotoxicology and environmental fate of the major chemicals used on potato crops enable comparisons to be made of the potential environmental impact of different pesticides, though they are a guide only and other factors may affect their behaviour in the field. The risk of a pesticide is assessed by measuring its toxicity and likely exposure in the field. Some are classified as poisons, including fentin hydroxide, chlorfenvinphos, phorate, aldicarb oxamyl, paraquat, and sulphuric acid.

A number of pesticides used on potatoes have been detected in groundwater in EU member states, including aldicarb, oxamyl (Netherlands), dimethoate (Denmark, UK), mancozeb, maneb, metalaxyl (Netherlands), linuron (UK), terbutryn (Netherlands, UK), terbuthylazine (Austria, Italy, Sweden), and tecnazene (UK).

Insecticides are more likely to cause environmental impacts than fungicides, because they kill non-target insect species as well as pests. Insecticides do vary in their selectivity however, for example, dimethoate has a very broad spectrum of activity, whilst pirimicarb is relatively selective. In terms of toxicity to other groups, insecticides vary too. Organophosphate insecticides tend to have relatively high mammalian and avian toxicity, whereas pyrethroids, typically exhibit low toxicity to mammals and birds, though they are highly toxic to fish.

The most environmentally damaging group of pesticides used on potatoes is the nematicides, particularly the most commonly used nematicide aldicarb, which is highly toxic to most animal life. Aldicarb is also highly soluble in water, and prone to leaching.

Steps to mitigate the use of pesticides on the environment generally exist already in formal pesticide registration procedures. In addition, some national assessment procedures have been developed to aid in relative impact of pesticides in the environment.

In The Netherlands, an environmental yardstick has been developed by the Centre for Agriculture and Environment to identify the relative toxicity of chemicals and their persistence in the environment. This yardstick considers: the potential for drift, toxicity to non-target organisms in the soil and water and the risk of groundwater contamination.

For each agro chemical, the potential risk is expressed in environmental impact points (EIP); the greater the score, the higher its impact on the environment. This system was developed in order to enable farmers to select pesticides with the least environmental impact and to quantify the environmental impact of the use of pesticides at farm, regional and national scales.

Water demand and irrigation

The potato crop is particularly responsive to water supply; conversely, it is drought susceptible in terms of yield and quality.

Assuming a yield of 25 t ha-1 and a volume of 1000 m^3 per ha water applied to irrigated potato crops, water use would be equivalent to 40 litres per kg potato production.

Irrigation is more important in southern Europe than in northern Member States, though irrigation is increasing in many northern Member States, where potatoes are the major irrigated crop. The method of irrigation and source of water varies across different areas of Europe. In the UK, potatoes are the most important irrigated crop. In England, 45% of the crop was irrigated in 1995 representing 60% of the area and 70% of the irrigation water volume used in England. In France, potatoes are not the main irrigated crop, with only 50,000 irrigated ha. In Spain, potato crops are generally irrigated. Only half the available area in Germany is irrigated, about 237,000 ha or 1.4% of the agricultural area. In the Netherlands, arable crops are rarely irrigated.

Starch potatoes are not irrigated and irrigation is uncommon in ware and processing potatoes, though irrigation is more common in seed potato crops.

In terms of estimating water requirement for the crop scheduling of irrigation based on soil moisture status and crop requirements can reduce water use without affecting yield. Trigger thresholds can be used to release first irrigation. Drip irrigation is more efficient in terms of water use than rain guns, but is expensive.

Abstraction of water may result in aquifer exhaustion and damage to rivers and wetland habitats. Yield remains the major motivation for irrigation, but irrigation at tuber initiation also helps to control common scab and therefore improve skin finish. Cultivars for processing are also more often grown under irrigated conditions in order to encourage uniform growth of tubers and uniform starch distribution inside the tubers.

In addition irrigation especially if applied unevenly, as large droplets or in excess of crop requirement can cause physical erosion and increased pollution. Soil erosion can cause pollution of watercourses by sediment. Potatoes are more susceptible to nitrate leaching than other crops, for a number of reasons. They are spring planted, leaving the land vulnerable to leaching over the previous winter.

Potatoes also have poor root systems, which are inefficient at taking up nutrients. Accordingly, soil nitrate residues in autumn are higher than after other crops. Potatoes are often grown on light soils in the UK and in The Netherlands. Such soils are more prone to leaching. Irrigation can improve efficiency of N uptake, but excessive irrigation can exacerbate leaching.

Nitrate and phosphate pollution of water leads to eutrophication. Nitrates affect mainly coastal waters, whereas phosphates are important in freshwater systems.

Movement of phosphorus occurs through erosion of soil particles and through dissolved and particulate suspension in run-off and groundwater. The potential for phosphate pollution of watercourses is therefore higher under potatoes than other crops because of the high phosphate inputs relative to other crops and because of the vulnerability of land under potato cultivation to erosion.

Organic potatoes: good or bad?

Interest in organic potatoes in the marketplace though tending to increase is variable. Poor skin finish in reduced input systems or true organic production, can lead to customer rejection. However, organic potato production should have a number of environmental advantages arising from the general avoidance of artificial inputs, the use of organic manures and rotations (which usually include grass-clover leys) for fertility building, and sympathetic management of non-crop habitats. However, some aspects of organic production may give rise to adverse environmental consequences.

The need for balanced rotations in organic farming systems and the exclusion of artificial fertilisers means that the position of the potato crop within the rotation is critical. Since potatoes require relatively high fertility status compared with other commonly grown organic crops, they are often planted after the fertility building ley (short term grass) period. Spring cultivations required to break down the structure of the grass ley to produce a tilth may result in an over fine tilth which is subject to erosion. However, production of tilth is less of a problem in the lighter stone-free soils used for potato growing in France, Belgium and The Netherlands, and if potatoes are grown after a cover crop, fewer cultivations are needed to prepare for potato planting.

Organic farming relies largely on organic manures to provide fertility. Farmyard manure can be a relatively inefficient method of supplying nutrients, and nutrients from liquid manure (slurry) may be used more efficiently. Even distribution of organics is difficult and can lead to local overdose of nutrients.

In Continental Europe, Colorado beetles are generally a greater problem than in conventional production systems. However there are several methods available for control available for organic farmers which do not have adverse environmental implications. Similarly, although insecticides cannot be used, alternative methods of limiting aphid attack to minimise virus infection of seed have been developed.
The main environmental problem arising from organic potato production is the use of copper-based fungicides to control Late Blight. Copper fungicides have higher toxicity to mammals and fish than the commonly used conventional fungicides, and are bound to soil particles so that they accumulate over time.

One criticism of organic potato crops is that they may act as a disease or pest source for traditionally grown crops. In this case, if substantiated, the production of organic crops could enhance pesticide use elsewhere!

Minimising the environmental impact of the potato

A number of codes of practices exist or are in preparation within Europe which set out standards for potato production practices. Some of these are general codes for protection of soil, water and air, some are explicitly concerned with "integrated crop production", and some are primarily concerned with maximizing the quality of the product, though minimising environmental impacts is also emphasized. In addition, several codes of practice exist for organic production, produced by the various bodies set up to regulate the organic sector.

The promotion of environmentally beneficial practices among potato producers can be achieved by a range of methods and through a number of types of organisation. Supermarkets have become more active in this area through the use of production protocols, but other sources of awareness or improved practices include processors, farmers unions, suppliers, consultants, advisors and peer pressure from other farmers. Information and advice offered in the form of software packages or via the Internet are also becoming increasingly available and more widely used.

Integrated production is a system of farming often referred to as an "integrated farming system (IFS)", involving a combination of cultural methods of pest and disease control, targeted agrochemical inputs and the use of new technologies, to reduce environmental impacts of farming. Integrated farming systems have received considerable publicity in recent years and are being adopted by an increasing number of farmers. Guidelines are now available in a number of EU member states for integrated production of potatoes.

Buffer zones reduce run-off and soil erosion, and also protect water courses from pesticide and fertiliser drift, and can also provide additional wildlife habitats. Although there may be costs from taking land out of production, these may be offset by using set-aside land or entering an agri-environment scheme. Conservation headland is a term used for not spraying the outer few metres of crop with insecticides or herbicides. The practice reduces the impact of pesticides on non-target organisms in the crop, and protects non-crop habitats. Originally developed for use in cereals, research work and on-farm trials in The Netherlands have shown that it may also be applicable to potatoes.

Codes of practice have been developed by Governments or their agencies (*e.g.* Codes of Practice for Protection of Soil, Water and Air, produced by the UK government), research organisations (*e.g.* the Dutch Centre for Agriculture and environment (CLM), the agricultural industry (*e.g.* UK Assured Produce Scheme, French CNIPT, German BOGK), or supermarkets (*e.g.* Albert Hein in the Netherlands, Tesco in the UK). In November 1999 a new initiative was launched by leading European food retailers (the EUREP group) to promote integrated production of vegetables and fruit. The resulting scheme is known as EUREPGAP.

Codes of Practice to minimise environmental impact of potato production have been drawn up by consulting existing codes of practice, extracting items relevant to the minimisation of environmental impacts, editing where necessary and incorporating with additions and modifications based on the authors own experience and expertise.

Conclusions

Undoubtedly the potato crop amongst several crops has potential negative impacts on the environment. Conversely, it is an important part of a balanced diet for humans and it production and processing aids employment particularly in the rural economy. In the future the steps already taken to minimize impact and match inputs to crop need should be continued and extended.

Authors note

This paper is written in a narrative style to impart an overview of the potato and the environment. Fuller details are available from and owe their origins to the document, “The Environmental Effects of Potato Production in the European Union”, which was prepared by DG Environment of EC by Nigel Boatman, Naomi Jones, Melvyn Askew (author of this paper) and Tina Swainston, with help from Alistair Leake, Paul Struik, Herwart Böhm, Enrique Ritter and Nathalie Verjux. The author is pleased to acknowledge the assistance of Miss Zara Court in the typing and final preparation of this paper and associated conference presentation.

Curriculum vitae – Melvyn Askew

Melvyn Askew is Head Of Agriculture and Rural Strategy at Central Science Laboratory of Defra based near York. He has over 40 years experience of growing and researching upon potatoes amongst many other crops and has considerable speaking experience on a worldwide basis. He was the UK national consultant for potatoes in UK for many years; Section Chairman of the Agronomy Section of EAPR for 6 years and a member of Council of EAPR for 6 years.

Company profile – Central Science Laboratory

Central Science Laboratory is the British Government laboratory based at York and underpinning agriculture, food and environment. This is a state of the art laboratory employing approximately 600 staff. It has a considerable portfolio for potato business especially in the potato disease; pesticide residues and agriculture/environment areas.

Sustainable agriculture and food chain: Status quo and perspectives for the potato food chain

L. Hövelmann[1], A. Schaffner[1], O. Christen[2] and F. Reinicke[2]
[1]German Agricultural Society (DLG e.V.), Department of Sustainable Agriculture, Eschborner Landstr. 122, D-60489 Frankfurt am Main, Germany
[2]Martin Luther University, Institute of Plant Production, Ludwig Wucherer Str. 2, 06108 Halle, Germany

Abstract

After publication of the AGENDA 21 by the United Nations in 1992, the concept of "Sustainable Development" was characterised intensively by normative approaches and largely shaped from the fields oriented to policy and society. However, Sustainable Development gradually also made its way into the world of business. In the agricultural and food branch, for example, economic data in business reports were supplemented by social aspects and reports on the environment situation, so that the documents were thus upgraded to become sustainability reports. This paper discusses Sustainable Development in the food chain against the background of growing relevance of the concept for business and society. Sustainability is increasingly maturing in the perception of the leading actors to become an innovative concept that assists in making economic development viable to face the future. However, this calls for comprehensible operational approaches. Considerations on the practicability of sustainability for farm and business management therefore focus on sustainability indicators. These measurement categories make it possible to steer an enterprise in achieving multiple sustainability objectives which are described by economic viability, environmental soundness and social compatibility. A concept for the development of sustainability indicators is shown and discussed against the background of the potato food chain. It becomes clear that sustainability will develop its full significance in the food chain if the principles of Sustainable Developments can be combined with the market signals, and these are essentially features of the product and process quality.

Keywords: sustainability, food chain, product and process quality, indicators

Introduction

At the latest since the UN Conference in Rio de Janeiro in 1992 with the final AGENDA 21 document, sustainability has arrived in the agricultural and food sector too. The document explained that enterprises, regions and national economies can only survive in the long term if their ecological impacts stand in a balanced relation to economic and social requirements. Accordingly this does away with the widespread mistake that sustainability is purely a question of environmental compatibility. However, it does not answer the question of how this important finding can be translated into daily business life, and an "AGENDA 21 process" was launched worldwide with a view to rectifying this vacuum. The provisional outcome of this process consists of a large number of good to well-meant approaches, some of which admittedly brought about genuine improvements, but whose greatest merit lies above all in a significantly enhanced awareness of and attention to the topic in business, politics and society.

Attempts have also been made to operationalise sustainability in the agricultural and food industry, but mainly without lasting success. The normative approaches and the complex policy concepts surrounding the term "Sustainable Development", in some cases only touching on niches, have even led to a certain weariness in some cases. This view has meanwhile given way to the finding that sustainability certainly possesses a high relevance for broad segments of business and society. After all, sustainability is increasingly maturing in the perception of the major actors to become an innovative basic framework that helps to make economic development viable for the future. For this, however, plausible, operational approaches are needed. Examples of these are described below with reference to the potato food chain.

Sustainability sub-objectives and categories

The concept of Sustainable Development distinguishes between sub-objectives (for example the economy), categories (for example profitability), and indicators (for example return on equity [%]). The prerequisite for dealing with sustainability is the availability of appropriate measurement categories and indicators within the categories. The indicators must be scientifically tenable and on a sector-specific level they should map the sub-objectives of sustainability precisely. Often several indicators belong to each category. These sub-objectives are described by economic, ecological and social aspects. Table 1 provides a simplified overview on the sub-objectives and the categories in the food chain.

Categories like liquidity, profitability and stability are suitable to describe the economic status of farms and of food processing industries. The potato food chain is an example for mainly integrated chains and therefore economic problems on one stage maybe influence the economic status on another stage, for instance when it comes to long term relationships in special quality programmes.

Table 1. Sustainability sub-objectives and categories in the food chain (simplified).

Sub-objective	Categories
Economy	Liquidity
	Profitability
	Stability
Ecology	Environmental impacts
	Consumption of natural resources
	Efficient use of resources
Social Matters	Employment
	Acceptance in society
	Self-realisation

Sustainability indicators

While the individual stages in the food chain only differ slightly from each other at the level of the categories, differences in composition can arise at the level of the indicators. Among the ecological indicators especially, differences are to be found between the farm-stage, that chiefly takes place outdoors, and for example the food industry, that generally produces in closed units, in other words in buildings, and thus uses the environmental areas of soil, water, air and biodiversity with different weightings.

Table 2 presents a choice of sustainability indicators within the different categories shown in Table 1. Some indicators can correspond to different indicators.

Ecological indicators are used to assess the influence of on the biotic and abiotic environment and to map the consumption of environmental goods (soil, water, air). One relevant assessment area in the category of environmental impacts is, for example, the eutrophication of waters by nutrients. A suitable indicator for this is the N balance [kg N/ha * a] or the nitrogen load [kg N/ha * a] within a year. As a borderline between sustainable and non-sustainable systems, a Nitrogen-surplus from 50 kg N/ha * year is discussed. The identification of borderlines is a challenging process between the different stakeholders in the food chain. Borderlines have to be revised from time to time to adjust to the actual state of the knowledge and technique.

Table 2. Sustainability categories and indicators (choice).

Ecology	Economy	Social Matters
Environmental impacts	Liquidity	Employment
• Erosion disposition	• Cash flow	• Remuneration for work
• Soil stress index	• Repayment of principal limit	• Upgrading of manager/ staff (areas of production technology, food safety, self-development)
• Proportion of ecological priority areas	• Ratio of outside capital	Acceptance in society
• Nutrient contents and supply of the soils	• Equity capital ratio	• Readiness to cooperate (membership in machine use groups, farming groups)
Consumption of natural resources	Profitability	• Honorary commitment
• Humus balance	• Calculatory results	• Use of management systems, participation in quality assurance measures, use of certifications
• Energy balance	• Contribution margin II	Self-realisation
• Crop rotation share of crops	• Equity capital profitability	• Days holiday manager/staff
Efficient use of resources	• Outside capital profitability	• Training activities
• N, P, K balance	Stability	
• Plant protection intensity	• Net investment	
	• Outside capital cover	
	• Profit rate	

In the potato food chain, the ecological sub-objectives can be described on the farm stage through indicators like "crop rotation share of potato" or "plant-protection-measurements". Both indicators have a strong linkage to potato quality aspects. On the stage of food processing industry ecological aspects can be described through "water-use [l/t]" or "energy-use [kJ/t]", each in reference to the produced units (*e.g.* potato-chips).

Another prerequisite for sustainable farming is its economic viability. This can be described very well in the categories liquidity, profitability, stability, and with ratios of the profit and loss account.

- Liquidity expresses the ability of the enterprise to meet payment obligations. It is the prerequisite for existence of the enterprise, as insufficient ability to pay leads to insolvency (DLG 1997, Heißenhuber 2000). One indicator of this, for example, is the equity capital ratio [%].
- Profitability provides information about the success and failure of entrepreneurial activities. It represents the ratio of profit to the production factors employed. (DLG 1997, Heißenhuber 2000). Example of an indicator: equity capital profitability [%].
- Stability describes the ability of the enterprise to secure its activity in the long term when unforeseen risks arise (DLG 1997, Heißenhuber 2000). Example of an indicator: profit rate [%].

The sub-objectives of sustainability in the social dimension consist, for example, of the appropriate stakeholding of the people working in an enterprise in the success of this enterprise. A sustainably managed enterprise should allow the people working there an appropriate income, grant them scope to act on their own responsibility, and become integrated in the social environment (Müller and Kächele 2000). On the basis of this definition, the three superordinate categories of employment, self-realisation and acceptance by societies are defined, from which indicators are to be derived. The social indicators thus characterise the working conditions and life situation of the persons working in the company. Furthermore, the indicators cover the social commitment of the manager and the possibility of personal development of management and staff.

A preliminary set of indicators can be derived from these requirements, but this must be specified more precisely in the still ongoing studies.

Certain quality requirements are demanded when it comes to selecting indicators. They should be relevant for decision-making, user-kindly, inexpensive to survey, based on high-quality statistical data, and be closely related to the location (v. Münchhausen and Nieberg 1997). The indicators can vary sectorally and depending on the level in the food chain, as it is the objectives and requirements pursued by a company that are sectorally crucial for selection of the indicators. That is why the selection of suitable indicators should be geared strongly to the purpose of use of the enterprise/actors (Wahmhoff 2003) and this is defined by internal and external stakeholders.

The purpose for which the indicators are used is defined at all stages of the food chain by internal and external stakeholders. Internal stakeholders, such as for example business proprietors, the management and staff, are interested in attaining the highest possible and secured income, increasing the value of the capital used, and enjoying good working conditions. External stakeholders, for example customers of the enterprise and consumers want safe and high-quality products at favourable prices and environmentally sound production, in other words primarily a high product and process quality (Odening and Bokelmann 2000). The concept of sustainability becomes operational through indicators, and as a management system makes it possible to optimise operations along various sub-objectives.

Sustainability and product- and process-quality in the potato chain

In 2003 the per capita consumption of potatoes was 67 kg. Of this amount 32.9 kg were consumed as fresh table potatoes and 34.1 kg as refined products (ZMP 2003). The potato chain is very heterogeneous as regards its degree of integration. The market is divided into highly integrated segments on the one hand, in which cultivation, processing and distribution are strongly interlinked organisationally, and very strongly subdivided segments on the other hand, with a large number of organisationally separate steps and intermediate stages.

Easily half of the potatoes produced in Germany reach the consumer as processed foods. Processed products comprise in particular dried and frozen products and potato crisps. The lion's share is accounted for by dried products, followed by frozen products and potato crisps.

The potatoes for processing are mainly cultivated under contract; potato processors are thus important partners in the market for farmers. The largest potato processors (starch potato processing is excluded) in Germany are:

- Stöver (processes 450,000 t/year);
- Bestfoods Pfanni (150,000 t/year);
- Mecklenburger Kartoffelveredelung (140,000 t/year);
- Nähr-Engel GmbH (100,000 t/year).

Sustainability on farms and in the food chain is very closely linked with questions of product- and process-quality. This reflects the strong relationship between sustainability and market requirements and concerns the competitiveness of the companies very directly.

For example, sustainability at the agricultural level can be described by fertilizer management (ecological dimension), the farmer's training status (social dimension) and the equity capital ratio/liquidity (economic dimension). The relationship between sustainability and issues of product- and process-quality now becomes clear from the fact that the fertilising has a major influence on the quality of the potatoes (internal and external quality parameters; ability of storage). The farmer's level of training is closely linked with his willingness or ability to implement hygiene management on the farm. The farm's liquidity reflects the long-term nature and reliability of the economic relations with the partners in the market, this too being a mark of the process quality.

In the potato chain different tools and documentation systems are used to ensure the required product and process quality of the agricultural raw materials, depending on the stage in the chain. One important instrument for communicating quality assurance and success is certification of the farms. A further instrument consists of cropping contracts, in which certain requirements made of the production are agreed between producer and processor. For example, subjects of such agreements could be organisation of crop rotation, soil-nutrient examination for needs-driven fertilising, the seed and plant material used. Furthermore, the production processes are to be documented. The parameters recorded include soil working and sowing, fertilising and plant protection measures, yield and quality results. Although sustainability is never explicitly measured or contractually agreed within the scope of these systems, very close relationships are evident.

Alongside these instruments there are directives and guidelines on sustainable production of potatoes and sustainable arable farming (SAI, EISA, EUREPGAP, QS).These schemes, some of which are connected with certifications, are to ensure higher qualities of the products produced and support the promotion of food safety (Table 3).

The systems in Table 3 aim on different targets and use different actions to reach the targets. They comprise for instance rules on cultivation measures, on the seed and plant material used, on hygiene management, on storage and transport of the harvested crop, on studies of crop residues, and on upgrading of the staff employed. Whether the criteria are satisfied is checked systematically in farm audits within the context of certification processes.

In addition there are also bilateral agreements between farmers and processors in the form of cropping agreements. These regulate, for example, supplier audits that support the selection of

Table 3. Systems, targets and actions to achieve quality and sustainability in agriculture.

System/Initiative	Targets	Actions
QS (shareholders: DBV, DRV, CMA, HfM)	Cross-stage securing of food safety	Definition of a product guideline with criteria of safe raw material production
EUREPGAP (members: METRO, EDEKA, GLOBUS, TEGUT)	Food safety, environmental protection at farm level	Definition of satisfaction criteria and control points for agricultural practice for benchmarking agricultural production. Continuous improvement in the areas of environmental protection, food safety
SAI (Sustainable Agriculture Initiative) (supporters: NESTLÈ, UNILEVER, DANONE)	Quantitative and qualitative safeguarding of agricultural raw materials, conservation of natural resources, consumer confidence	Development of principles and practices of sustainable farming (for example for potatoes, grain)
EISA	High-quality raw materials, environmentally sound production, conservation of natural resources taking society's requirements into account	Recommendations on sustainable operating in the fields of management, monitoring, plant protection, use of energy, water and soil protection; preparation of best practice examples

(supporters: FNL, LEAF, FARRE)

suppliers on the basis of criteria that can be validated. There is a particularly strong link between farmer and processor in the "closed systems". Here the food-producing company intervenes in detail in the farm unit and regulates for example details of seed propagation, or steers other production measures, such as for example plant protection via in-house consultants. In contract cropping some producers use their own database systems and capture the agricultural production data themselves. In this way data on the plant material, cropping processes and the quality of the raw materials produced are recorded. Furthermore, the processing cycles in the factory "right through to the crisps bag" are documented. This documentation allows traceability right back to the field.

A common factor of the various forms of quality assurance is that the impacts of the management measures, for example on the environment, are not quantified. These criteria are generally based on a "satisfied - not satisfied" check.

What benefit do farmers expect from implementing sustainability in the food chain?

The benefit of sustainability indicator systems is seen in particular in improved market access for the agricultural products, in optimising in-house cycles, and in improving operational documentation (Hövelmann 2003).

These findings by experts at a sustainability hearing in spring 2003 were essentially confirmed by farmers in current focus group surveys (Table 4). The customary documentation procedures on the farms were determined and the expected benefits linked to a possible sustainability documentation were analysed. The interviewees distinguished between internal and external evaluation and documentation options. The internal evaluations (for example field file, annual accounts and analysis of the production branch) assess the management results and serve as a basis for the further production-related decisions.

The external evaluation options cover the documentation duties vis-à-vis public authorities and customers and should be tailored to the respective obligations to furnish evidence. The current customary records of the farmers serve to satisfy the state documentation duties and the requirements of business partners. Records and evidence extending beyond current duties are often seen as an extra burden and viewed sceptically. However, the extra outlay is accepted if higher producer prices or safer sales of the products produced are achieved. Sustainable production measures thus depend on the demand for sustainable raw materials in the value chain.

Users also find flexible evaluation options, freely definable data export, an option for horizontal farm comparisons, and legal and data security important. Interviewees generally rejected Internet-based systems, as secured data protection appeared questionable.

Sustainability: What do farmers say about it?

In view of its central position in the value chain, the agriculture stage assumes a leading role with regard to sustainability. That is why it is extremely interesting to see what farmers

Table 4. Expected benefits and statements from farmers about sustainability-management systems in focus-groups surveys in 2005.

Expectations	**Statements/Explanations**
Benefit for management	Farm sustainability analysis and sustainability certification are accepted if: marketing channels are secured, there are advantages in loan granting and leases, and the danger of product defects is reduced.
	The systems must support the farmer in completing documentation to meet legal requirements (satisfying duties to furnish evidence)
Evaluation options	It must be possible to compile the evaluations as required by the target groups to allow them to furnish necessary evidence to public authorities and their many business partners.
Compatibility with other systems	Data duplications are to be avoided to reduce work input
Data security/confidentiality of data keeping	Central, web-based systems are viewed sceptically as regards data confidentiality

feel about this topic. In the current DLG-Trendmonitor 2005, altogether 145 farmers were questioned about their attitude to sustainability (Figure 1).

Some 56 percent viewed the term positively and thought it was a good designation. However, almost 43 percent thought that the name was too diffuse and complained of a lack of a transparent and comprehensible definition. Only 5.7 percent of the farmers thought that sustainability had negative connotations. Ten percent of the farmers offered important cues

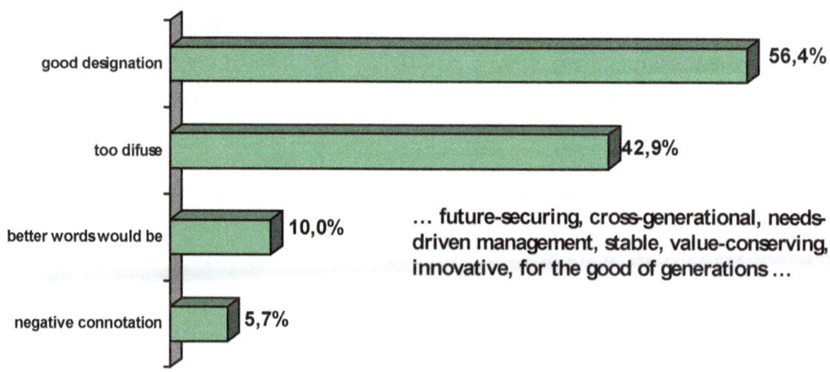

Figure 1. What farmers think about sustainability. Source: DLG-Trendmonitor 2005; n=145; arable farms in Germany.

for a better conceptuality: future-securing, cross-generational, value-conserving, innovative. And this addresses precisely those objectives that are connected with Sustainable Development in all leading national and international publications.

Prospects

The combination of sustainability on the one hand and product- and process-quality on the other offers great opportunities for the food chain altogether - especially when sustainability is not grasped as an end in itself - but instead is harnessed to foster the stability and forward viability of economic enterprises. The linking of sustainability and quality parameters joins socially highly relevant themes, such as environmental and social compatibility, with the clear signals from the markets in which the food chain enterprises move.
At present discussions on sustainability in the various food chains still concentrate very strongly on the agricultural level. However, in order to reinforce Sustainable Development in the entire food chain, it is necessary for the individual links in the chains to harmonise more strongly with each other. Points of approach for this can be seen in the growing activities to increase transparency and traceability in the chain. The technology of precision farming with its enormous variety of technical and organisational elements also offers major potential for optimising the food chain in accordance with the multiple goals of sustainability.

Acknowledgements

The authors wish to thank the German Federal Ministry for Education and Research (BMBF) (joint research project preagro II) and the German Federal Environmental Foundation Deutsche Bundesstiftung Umwelt (DBU) (Ref. 22544) for supporting the research work.

References

Deutsche Landwirtschafts-Gesellschaft (DLG) (1997): Effiziente Jahresabschlussanalyse. Neue, einheitliche Erfolgskennzahlen für landwirtschaftliche Betriebe aller Rechtsformen. DLG-Verlag, Frankfurt am Main.

Heißenhuber, A. (2000): Nachhaltige Landbewirtschaftung - Anforderungen und Kriterien aus wirtschaftlicher Sicht. In: VDLUFA Kongressband 2000 Generalthema „Nachhaltige Landwirtschaft", 112th VDLUFA Congress

Hövelmann, L. (2003): Arbeitsgruppe Nachhaltigkeit in der Landwirtschaft - messen, bewerten, managen. In: Girnau, Hövelmann, Wahmhoff, Wolf, Wurl (Eds.): Nachhaltige Agrar- und Ernährungswirtschaft. Herausforderungen and Chancen in der Wertschöpfungskette. Erich Schmidt Verlag, Berlin

Münchhausen H. v. and Nieberg, H. (1997): Agrar-Umweltindikatoren: Grundlagen, Verwendungsmöglichkeiten und Ergebnisse einer Expertenbefragung. In: Umweltverträgliche Pflanzenproduktion - Indikatoren, Bilanzierungsansätze und ihre Einbindung in Ökobilanzen. Deutsche Bundesstiftung Umwelt

Odening, M. and Bokelmann, W. (2000): Agrarmanagement. Ulmer, Stuttgart, 372 p.

Wahmhoff, W. (2003): Nachhaltigkeit managen mit Indikatoren. In: Girnau, M., Hövelmann, L., Wahmhoff, W., Wolf, W., Wurl, H. (Eds.): Nachhaltige Agrar- and Ernährungswirtschaft. Herausforderungen und Chancen in der Wertschöpfungskette. Erich Schmidt Verlag, Berlin

ZMP (2003): Marktbilanz Kartoffeln 2003. Zentrale Markt- und Preisberichtsstelle für Erzeugnisse der Land-, Forst-, and Ernährungswirtschaft, Bonn.

Curriculum vitae – Dr. Lothar Hövelmann

Dr. Lothar Hövelmann was born in 1962 in Duisburg, Germany. Following his apprenticeship on three farms in North-West Germany he studied agricultural sciences at Bonn University. After obtaining his degree in 1990 he worked as an agricultural consultant at STEAG AG Essen. There he advised the company on changing the cultivating system of its 300 ha farm from conventional to organic farming.

In parallel, during this period, he started his field experiments at Bonn University within the framework of his PhD thesis. He experimented with reduced tillage systems and their effects on physical soil properties. In 1993 he moved to GRUBE Land- und Umwelttechnik in Brake, Unterweser. There he was responsible for the research and development activities of the company and for managing the company's bio-waste composting plants. His scientific work during this period focused on developing defined organic-mineral fertilisers based on organic municipal solid waste and their field application.

Since 1998 Dr. Hövelmann has been employed by the German Agricultural Society (DLG e.V.). He is Head of the Department of Sustainable Agriculture and General Manager of the umbrella organisation of German scientific societies in the field of agricultural and environmental sciences (Dachverband Agrarforschung DAF e.V.). The focus of his work is now on research and development in the field of sustainable agriculture and he is responsible for the technical programme of AGRITECHNICA.

He is member of the steering committee of DFOR (German Forum on Development Oriented Research), designated member of the advisory committee of VdLUFA, and a member of various committees *e.g.* Verband der Landwirtschaftskammern (VLK), Fördergemeinschaft Nachhaltige Landwirtschaft (FNL) and CMA.

Company profile – German Agricultural Society (DLG)

The German Agricultural Society (Deutsche Landwirtschafts-Gesellschaft, DLG) was founded in 1885 by the engineer and author Max Eyth. With over 16,000 members it is one of the leading organisations in the German agricultural and food sector. The DLG is a specialist organisation open for anybody to join. It is politically and economically independent. The DLG's mandate and understanding of its work is to promote technical and scientific progress. With its activities and initiatives the DLG sets standards and provides impetus for progress. The DLG thinks and acts in international terms. It conducts knowledge sharing world-wide with leading international practitioners and other specialist organisations. Activity areas are: identifying solutions for the future, providing market overviews, testing products and issuing awards. Some 200 full-time staff and more than 3,000 honorary experts jointly work out solutions for topical problems. The over 80 committees, work groups and commissions form the foundation for expertise and continuity in the DLG's technical work. The DLG discusses, develops and communicates solutions for future challenges at over 40 conferences and seminars every year. The DLG seminars and its trainee programme create key qualifications for the future. With its many public sessions, the DLG Winter Conference as a forum for the future sets major signposts for the agricultural and food sectors. The DLG Farm Management Conferences take up topical themes and present a discussion forum for leading practitioners. The DLG promotes the product quality of foods on the basis of impartial and binding quality standards. Its quality understanding is based on current scientific findings and self-developed, recognised methods. With its international competitions the DLG is Europe's leader in quality assessment. DLG exhibitions such as Agritechnica, EuroTier, Anuga FoodTec and DLG Field Days are the platform to present innovations, modern products, processes and services, and to promote intensive dialogue.

French quality and environmental schemes: Meeting customer requirements- the potato sector's approach

A. de Montigny[1] and N. Verjux[2]
[1]CNIPT Comité National Interprofessionnel de la Pomme de Terre, 9 rue d'Athènes, 75009 Paris, France
[2]ITPT-ARVALIS-Institut du végétal, Station d'expérimentation, 91720 Boigneville, France

Abstract

In France, many farmers are involved in a number of agricultural and environmental schemes, through the development of private accreditations or contracts and national or international codes of good agricultural practices. This abundance of schemes leads to multiple inspections by second (customers) or third (accreditation bodies) parties. Various options are available to reduce the constraints on farmers. Among those solutions is an attempt at implementing national production guidelines, which form the basis for additional quality specifications and other quality assurance schemes. In the potato sector, a French standard was launched in 2001 and is now implemented by half of the fresh market.

Keywords: quality and environmental schemes, potato standard

Introduction

Certification of products or farms, integrated farming, sector specific good practice charters, buyers' specifications, Agriconfiance®. The quality scheme scene in France may seem somewhat bewildering at first. In fact, different schemes suit different issues whilst often complementing each other. The main problem lies in the large number of inspections for farms involved in several schemes simultaneously. This article first outlines the differences between schemes, then shows how national good practices charters can help standardise the methods of production that must be implemented, without inhibiting competitive differentiation and freedom of decision for the farmer. Finally, it examines in detail how the French potato sector developed a now widely recognised production standard.

The quality scheme scene in France

French agriculture is controlled by a huge number of quality and environmental schemes. One of the keys to differentiating between them involves answering a simple question: does the scheme involve the farm as a whole, in which case it is referred to as a farm-based scheme, or is it dedicated to a specific product or group of products, which is then referred to as a product-based scheme. (Verjux and Bousquet, 2004).

For farm-based schemes, the main aim is to answer meeting societal expectations; in that case, the environment, safety and employee's rights will be given priority. They deal with issues such as integrated farming certification, a recent scheme launched by the French ministry of agriculture. They may also involve improving the quality of internal organisation, with ISO 9001 or 14001 accreditation, which have seen a low level of uptake amongst French farms. Such schemes do not include objectives directly linked to the quality of what the farm produces, but are concerned with the quality of the equipment, of the organisation of agricultural activity and of the farmer's overall competence.

The purpose of product-based schemes is primarily to meet direct buyers' expectations. The criteria used to assess product quality, are therefore essential in this type of scheme, which will focus on defining the production methods needing to be implemented to fulfil the required quality criteria. However, since direct buyers echo societal expectations through a succession of specifications set out down the production chain, some of the requirements expressed by the general public are often reflected in the conditions set out by direct clients. Environmental risks illustrate this point and are now widely included in product-based quality schemes.

Combining several quality schemes on a farm: need for national codes of good practices

The combination of several schemes on one farm can result in considerable obligations, firstly with the registration process and then the inspections. With a view to making things simpler and avoiding unreasonable duplication, the main French agricultural sectors have established national codes of good practices which are now used as the basis for other quality and environmental schemes (Verjux, 2000 ; Verjux and Veron-Delor, 2001).

In the crop sector, this type of approach currently involves cereals, maize, pulses and ware potatoes destined to the fresh market (table potatoes) (see Table 1). Those codes include issues such as technical and safety quality, environment and traceability (Arvalis, Irtac, Unip, 2004 and 2006). They are written by interprofessional working groups, tested before being circulated to producers and regularly updated. Compliance may be implemented differently depending on the sector concerned, but independent third party audits establish their credibility in all cases.

Product-based schemes fulfil two main purposes as far as clients are concerned:
- they bring the sort of guarantees linked to best practices as they take into account technical as well as societal expectations, and regulatory requirements,
- because of their national status, they are recognised by or transferable to other schemes, avoiding therefore duplication of initiatives and multiple inspections for the same crop.

Indeed, matching the contents of private specifications, quality assurance schemes and production standards help move towards standardisation of technical contents and make it possible for one inspection to be used for several schemes (Maze *et al,* 2004).

As we will see further on, all major French distributors and supermarkets use the NFV 25-111 standard for production of ware potatoes for the fresh market. As far as cereals are concerned, quality assurance schemes are used as a basis for specifications issued by processors (Malteurop, Association de la Meunerie Française) and for officially accredited cereal products. More than half of the total French malting barley production is involved in the scheme (Verjux, 2003 a and b).

Table 1. Quality Assurance Schemes and production standards for French crops.

	Quality charters for cereals	**Quality charter for pulses**	**Potato standard NFV25-111**
Crop involved	Soft wheat Durum wheat Malting barley Forage-grain maize Sweet corn	Peas Faba beans	Ware potatoes for the fresh market
Uptake level	70 groups representing 20,000 producers in 2005 ½ of the malting barley French production	Being launched in 2006	1400 producers, 1/2 of the French fresh market production
Compliance method	Group	Group	Individual or group
List of groups involved	www.arvalisinstitutduvegetal.fr		
Contact	Arvalis-Institut du végétal/ IRTAC	Arvalis-Institut du végétal/UNIP	CNIPT

The charters are also used as a basis for the specifications imposed by some Agriconfiance® certifications. Agriconfiance® is a scheme launched by French cooperatives to help build relations between the cooperative and its members. Finally, recognition of equivalence has been obtained for some private schemes established by processors (Socle McDonald's France, AGRIXO from Grands Moulins de Paris).

All these examples bring us closer to standardising technical commitments between one scheme and another and to one single inspection being valid for several schemes.

Codes of good practices are also developing at international level. Provided their promoters are willing to take into account existing initiatives, it will be possible to have national, sector specific schemes recognised as equivalent. This exercise has just been undertaken for French cereal quality assurance schemes, as part of a platform for sustainable farming (SAI Platform) established by food processing companies (DANONE, UNILEVER and NESTLE originally). Recognition was recently obtained for the French sweet corn quality assurance scheme, and its associated farm-based scheme, as part of the EUREPGAP standard for fruit and vegetables.

In 2005, some of those quality assurance schemes launched trade marks that can be used for finished products. Agreements entered into with processors help guarantee the traceability of each batch from the producer to the various organisms involved in the chain. The logo is accompanied by a short text describing the scheme implemented on the ground. This is a simple and practical way for the farmer committed to the scheme to reach consumers and get everyone along the chain interested in the specifics of agricultural production, including in the case of sectors involving a long chain (Verjux, 2006). Other sectors should follow suit with similar logos and the first products bearing the stamp should reach the shelves in 2006.

Developing a French potato standard

The CNIPT (Comité National Interprofessionnel de la Pomme de Terre), which brings together the whole Fresh Potato Sector - producers, cooperatives, packers, retailers -, undertook as early as in 1998, to draw up a code of good practice aiming to achieve a triple objective:

- to better take into account the environmental impact,
- to guarantee production competitiveness,
- to control certain of the qualitative characteristics requested by the market, by going beyond the regulatory requirements on some points: definition of nitrate content, minimum storage temperature of 4.5°C (to minimise sugar content and maintain nutritional levels) and chlorpropham content below a 5 mg/kg threshold.

The clarification and definition of the conditions for integrated production of potatoes by means of standardisation, expresses the will of everyone involved in the French sector, from production to distribution, to meet the expectations of the market which reflects consumer demands.

At first, a technical guide was produced with the expertise of ITPT/Arvalis-Institut du Végétal, together with the active collaboration of INRA (Institut National de la Recherche Agronomique) and of the Plant Protection Department (Services de la Protection des Végétaux) (CNIPT *et al*, 2000). This document defines all of the components of ware potato production and crop management, from planting to storage and packaging, with regards to environmental protection, production competitiveness and control of the technological and safety quality of the product.

This approach relies on tools, methods and models which guide the farmer and the store keeper in their decision-making. Each procedure is therefore reasoned and optimised on the basis of crop needs and harvest expectations of the environmental impact and of health risks.
With the same objectives and the same methods, other sectors (wheat, durum wheat, maize, sweet corn, malting barley, *etc.*) have undertaken similar concerted approaches as we have seen before.

An approved French standard: NF V 25-111

Added to the focus on traceability prevailing throughout the sector and to the current standardisation of delivery acceptance conditions (AFNOR, 2001), this standard will help more specifically to improve transparency in the potato market, especially as it is officially approved by the French Authorities, consumer groups, as well as other interested parties, such as growers and retailers (AFNOR, 2000). This document applies to ware potatoes intended for the fresh produce market and defines all the components of ware potato production and crop management, from planting to storage and packaging, in accordance with the objectives of integrated farming.

Moreover, production methods complying with other requirements may only invoke compliance with this document if those additional requirements are compatible with each of the three objectives defined in the introduction: competitiveness, taking into account the environmental impact and control of the qualitative characteristics of the product (technological and safety quality). For each following chapter, the points of compliance are described:

- establishment of the crop;
- choice of field;
- choice of varieties and preparation of plants;
- planting;
- fertilisation;
- organic manures;
- P, K and Nitrogen fertilisation;
- crop protection against soil and vegetation borne pests, waste, late blight;
- irrigation;
- harvest management (haulm destruction, lifting);
- storage and conservation management and preparation of delivery;
- marketing;
- management and apalication of fertilisers (storage, spreading equipment, spreading);
- management and application of chemicals (storage, sprayer and spraying);
- transport of fertilisers and chemicals by the farmer;
- discarding packaging and unused products;
- records;
- internal quality audit;
- labelling.

This standard was officially published in 2000 as NF V 25-111 by the French institute of standardisation AFNOR.

Development of the French standard

A review of the standard is beginning this year, in order to take into account societal changes and consumers' requirements, in particular with regards to hygiene and safety. This NF V 25-111 standard is the basis of the code of good practice for the table potato sector, and is now implemented by about 50% of the French market: even if this percentage is insufficient and due to its stringent requirements, it is a targeted response to many buyer schemes and protocols of the retail sector. As it is easily auditable according to EN 45011, the auditors may combine the audit with others required by the buyer.

Conclusion

In addition to French regulations, a farmer may be in a position where he needs to manage 2 or 3 sets of references per product on his farm. This once multiplied by the number of products and activities can quickly become unmanageable in terms of technical management and inspections.

National schemes developed for crops have helped standardise the technical commitment required by customers and limit duplication, without suppressing specific characteristics linked to each market. The level of recognition already achieved has rationalised the farm inspection process, but it must be further improved at an international level in order to achieve maximum rationalisation, without undermining consumer and environmental protection.

References

AFNOR (2000). NF V 25-111, bonnes pratiques de production de la pomme de terre de conservation destinée au marché du frais, compatibles avec les objectifs de l'agriculture raisonnée (good practice for production of ware potatoes intended for fresh produce market, compatible with reasoned agriculture), October 2000.

AFNOR (2001). NF V 25-112. Pomme de terre, bonnes pratiques de la réception pour la pomme de terre de conservation destinée au marché du frais (good practice for delivery acceptance of ware potato intended for the fresh produce market), Mai 2001.

ARVALIS-Institut du végétal, IRTAC (2006). Charte de production du maïs grain et du maïs fourrage, charte de production du maïs doux (maize and sweet corn production charters), March 2006.

ARVALIS-Institut du végetal, IRTAC (2004). Chartes de production du blé tendre, du blé dur et de l'orge de brasserie (wheat, durum wheat and malting barley charter of production), September 2004.

ARVALIS-Institut du végétal, UNIP (2006). Charte de production des protéagineux (pulses charter of production), January 2006.

CNIPT, ITCF, ITPT (2000). Guide de production raisonnée de la pomme de terre, marché du frais (guideline for the ware potato production for the fresh market), March 2000.

Maze A., Aït-Assa Myriam, Verjux N., Carrotte G., Hedouin C., Vaucelle A. (2004). Audit regulation and the improvement of benchmarking procedures among quality assurance schemes in agriculture. 92nd European Seminar of the EAAE, 2-4 March 2004, Göttingen, Germany.

Verjux N. (2000). Development of integrated production of malting barley in France. In Proceedings of international congress Innovation in the barley-malt-beer chain. 16-17 October 2000, Nancy, France, page 14 to 22.

Verjux N. (2003a). Qualitätsicherungs-/Umweltschutzmassnahmen bei den Produktzweigen des Ackerbaus in Frankreich. Internationale Ackerbautagung Qualitäts-Sicherungsstrategien im Ackerbau. 11 November 2003, Hannover, Germany.

Verjux N. (2003b). Chartes de production ARVALIS-Institut du végétal - IRTAC (Charters of production). In proceedings of malting barley seminar « un atout pour la France » (an asset for France), 5 February 2003, Paris, France, pages 21 to 28.

Verjux N. (2006). Valoriser ses pratiques : quelles opportunités en grandes cultures ?(add value to agricultural practices : which opportunities for crops ?) Perspectives agricoles, n°324, June 2006, pages 12 to 15.

Verjux N., Bousquet N. (2004). Démarches qualité en grandes cultures, des initiatives à découvrir (quality schemes for crops, a view of initiatives). Editions ARVALIS-Institut du vegetal, September 2004.

Verjux N., Veron-Delor G. (2001). Orge de brasserie française : la qualité dans toutes ses composantes (french malting barley, all components of quality). BIOS International, n°4, 2nd semester 2001.

Curriculum vitae – Aymard de Montigny

Aymard de Montigny is Maître de Sciences Naturelles, Technical & Quality Manager at the Comité National Interprofessionnel de la Pomme de Terre (CNIPT) - National board of the French table potatoes' industry, from the grower to the consumer.

He worked at Rhône-Poulenc Company in marketing pesticides from 1970 to 1979. He has been involved in product quality, communication and consumer awareness since 1979. He is the coordinator for the applied research and has contributed to the development of the quality, traceability from the grower to the supermarket, by standards, methods, controls.

Curriculum vitae – Nathalie Verjux

Nathalie Verjux is a senior scientist of ARVALIS Institut du Végétal for all aspects related to quality and environmental schemes or traceability in agriculture. As such, she has been a major contributor to the building of the French charters and she is now in charge of the updating, implementation and valorization of these codes in cereals sectors. She is member in a range of working groups in France dealing with traceability or quality assurance schemes (official member of CNAR-Integrated farming Committee, AFNOR, certification body,...). She coordinates the research program of the institute in different topics concerning good agricultural codes of practices (hygiene package, biofuels, ...) and traceability at the farm level. She is the head of environmental and cropping systems service (8 engineers). Before 2000, she was involved in long term fields experiments to evaluate different cropping systems approaches (low input, no-ploughing, decreasing time spent for the crop, *etc.*).

Company profile – Comité National Interprofessionel de la Pomme de Terre (CNIPT)

Comité National Interprofessionnel de la Pomme de Terre (CNIPT) is the national board of the French table potatoes' industry, from the grower to the consumer. CNIPT is involved in product quality, communication and consumer awareness.

Company profile – ARVALIS

ARVALIS - Institut du végétal is an agricultural applied research French organisation financed and managed by producers. ARVALIS' mission is to provide useful, usable and used information, techniques and services to farmers, agricultural organizations and firms from the various sectors, for them to adapt their production to the market changes using environment-friendly cropping systems. Within the framework of this mission ARVALIS contributes to strengthening a sustainable agriculture using all the technologies of an integrated, precise and innovating agriculture for:

- Improving the competitiveness of farming businesses.
- Continuously adapting products quality to the actual and potential demand of European and third countries markets.
- Controlling agricultural activities impact on the environment.
- Providing objective arguments promoting agriculture image.

ARVALIS' research concerns cereals (soft wheat, durum wheat, barley triticale, rye, oats, rice...), maize (grain, forage, seed, sweet corn), sorghum, protein crops (peas, faba beans,...), potatoes and forage crops.

With its research stations, trial programmes, implementation farms and laboratories, the Institute has 35 sites located throughout France, and 400 collaborators, 300 of which are scientists or technicians. They lead studies in fields such as agronomy, crops science, biotechnologies, fertilisation and crop protection, precision farming, harvest and storage, quality, markets and economics of the different sectors as well as of farming businesses.

Agronomic strategies for the organic cultivation of potatoes for processing into high quality French fries and potato crisps

H. Böhm[1], T. Krause[1], T. Haase[2], N.U. Haase[3], R. Loges[4] and J. Heß[2]
[1]Federal Agricultural Research Centre, Institute of Organic Farming, Trenthorst 32, 23847 Westerau, Germany
[2]University of Kassel, Department of Organic Farming and Cropping, Hessian State Domain, D-34393 Grebenstein, Germany
[3]Federal Research Centre for Nutrition and Food, Institute for Cereal, Potato and Starch Technology, Schützenberg 12, 32756 Detmold, Germany
[4]University of Kiel, Institute of Crop Science and Breeding, Hermann-Rodewald-Str. 9, 24098 Kiel, Germany

Abstract

In order to achieve the successful cultivation of potato raw material to be processed into French fries and crisps, the high quality standards set by the processing industry have to be fulfilled and high tuber yields attained. In a cooperative research project, different strategies for the cultivation of high quality potato raw material for processing were examined under the conditions of organic farming. The choice of variety was found to be the most important agronomic instrument for the organic farmer to secure high tuber yields and quality. Varieties Agria and Marena can be recommended as they gave high tuber yields of the required size grade and also high French fry quality both at harvest and after storage. With regard to processing into crisps, cv. Marlen is highly suitable because it was observed to have good quality of raw material even after storage. On the other hand the level of reducing sugars in tubers exceeded the maximum tolerable concentration in relation to the maturity stage at lifting. Furthermore, results confirm that any available agronomic measure to improve nutrient and water availability, but also seed-tuber treatment, has to be exploited in order to secure high yields at a satisfying quality level. Leguminous crops (field peas (*Pisum sativum* L.) with a following catch crop and ryegrass-clover leys) can be recommended as preceding crops. The management of ryegrass-clover leys (cutting or mulching) does not have a significant impact on tuber yield, grading or quality. Agronomic measures did influence processing quality, yet the impact of the weather conditions on crop development was much more important. In summary, it can be stated that organic potatoes for processing can be cultivated successfully when site- and farm-specific conditions are considered.

Keywords: organic farming, French fries, potato crisps, yield, processing, quality

Introduction

The processing industry sets high standards for processing potatoes with regard to their internal and external quality. Minimum concentrations of starch, or respectively dry matter, are important as are low concentrations of reducing sugars.

Potato growers interested in producing raw material for crisps or French fries production should consider the varieties best suited for industry, as well as varieties that produce well under their growing conditions. A comparatively high concentration of dry matter (French fries 19-23%, crisps 21-25%), or rather starch (French fries 14-18%, crisps 16-20%), is necessary, as the concentration of reducing sugars (glucose and fructose) should not exceed the benchmark of 150 mg 100 g^{-1} (crisps), resp. 300 mg 100 g^{-1} in fresh matter (French fries). Furthermore, the processing industry requires tubers with a good suitability for storage at 8°C and a high proportion of tubers over 50 mm (French fries).

Hardly any knowledge exists about quality of French fries and crisps from organically cultivated potato varieties (Böhm *et al.*, 2002). Based on the first experiences of Böhm *et al.* (2002), field experiments with selected potato varieties appropriate for potato crisps and French fries processing were examined in a two year study at well-defined locations.

Due to the limited possibilities to ensure nitrogen supply with fertilisation in the system of organic farming, the crop preceding potatoes is of major interest. The nitrogen supply is supposed to be influenced by different previous crops like ryegrass-clover mixtures, pulses and cereal grains. In a two-year field experiment, the influence of different previous crops was examined with regard to yield and quality on the experimental farm of the University of Kassel. Furthermore, experiments by Loges (1998) showed that the efficiency of nitrogen fixation depends on the management system (mulching or cut mowing). Also the nitrogen mineralisation can be affected by different C/N-conditions of ryegrass-clover residues on the field. Over a period of two years (2003/2004), the effect of different ryegrass-clover management systems on yield, grading and quality of processing potatoes were examined on the experimental farm of the University of Kiel in northern Germany.

Methods

Variety trial

An assortment of eight potato varieties, all appropriate for processing into potato crisps or French fries according to the German Certification Office, was cultivated on three different sites in field experiments with four replicates under the conditions of organic farming in the years 2003 and 2004. Only in Wulksfelde the variety trial was irrigated with 120 mm in 2003. In Table 1 the sites and important cultivation facts are listed.

Previous crop management experiment

In Frankenhausen (central Germany, Hesse), a field experiment with different previous crops (alfalfa ryegrass-clover mixture, field peas (*Pisum sativum* L.), oat (*Avena sativa* L.) and winter wheat (*Triticum aestivum* L.) was conducted in the years 2003 and 2004 (Table 2). Furthermore the factor "pre-sprouting" was included in this trial, but shall not be discussed in this paper. So, the average of the pre-sprouted and non-pre-sprouted plots are shown.

Clover grass management experiment

At the location Lindhof (northern Germany, Schleswig-Holstein) another two-year field experiment was carried out with different ryegrass-clover management systems. In spring, the first tillage operation of the ryegrass-clover grass was done by a rotavator. Afterwards, before

Table 1. Characterisation of sites and cultivation of the variety trials in the years 2003 and 2004.

Site	**Lindhof (LI)**		**Wulksfelde (WU)**		**Frankenhausen (FH)**	
Location	Northern Germany, Kiel		Northern Germany, Hamburg		Central Germany, Kassel	
Soil type	Sandy loam		Loamy sand		Clay silt	
Year	**2003**	**2004**	**2003**	**2004**	**2003**	**2004**
Preceding crop	Oat	Oat	Ryegrass-clover	Ryegrass-clover	Rye	Wheat
Catch crop	Mustard	Mustard	-	-	Blue lupine/ Phacelia	Blue lupine/ Phacelia
Pre-sprouted	yes	yes	yes	yes	yes	yes
Date of plant	22. Apr	26. Apr	16. Apr	21. Apr	16. Apr	6./16. Apr
Date of harvest	17. Sep	16. Sep	10. Sep	7. Sep	10./17. Sep	21./23. Aug 10./11. Sep

Table 2. Characterisation of sites and experimental factors in the years 2003 and 2004.

Experiment	**Previous crop management**	**Ryegrass-clover management**
Year	2003/2004	2003/2004
Location	Central Germany, Kassel	Northern Germany, Kiel
Site	Frankenhausen	Lindhof
Soil type	Clay silt	Sandy loam
Date of planting	24. Apr 03 / 22. Apr 04	22. Apr 03 / 28. Apr 04
Date of harvest	17. Sep 03 / 18. Sep 04	17. Sep 03 / 15. Sep 04
Factor I (plot)	Management a) Alfalfa ryegrass-clover b) Field pea c) Winter wheat d) Oat	Management a) 3x cut mowing b) cut mowing/mulching c) 3x mulching
Factor II (subplot)	Potato variety a) Agria (French fries) b) Marlen (crisps)	Potato variety a) Agria (French fries) b) Marlen (crisps)

planting the pre-sprouted potatoes, the soil cultivation was conducted using the soil separation technique (Böhm, 2002).

Table 2 characterises the sites and sums up the experimental factors. If necessary haulm killing was done by a haulm cutter two weeks before harvest. Further cultivation of the experimental fields occurred under customary circumstances. The experiments were designed as a split-plot system with four replicates. Main plot treatment was the previous crops, resp. -the ryegrass-clover management, sub-plots consisted of the two varieties Agria (French fries) and Marlen (potato crisps). Precrops were ploughed under at the end of January.

In all field experiments, total tuber yield and the portion of different size-grades were assessed. Moreover tuber concentration of dry matter and starch (both measured by underwater weight) as well as reducing sugars (enzymatical method) according to Boehringer (1995) were determined as parameters of quality. In addition, an aliquot from each plot was processed towards potato crisps and/or French fries. The quality was analysed both after harvest and after a four-month period of storage at 8°C (exception: storage of the so-called 4°C varieties (Putz, 2004) Sempra and Verdi at 4°C).

Statistical analysis was done applying a mixed model (PROC MIXED) using the SAS package (8.2) (Littell *et al.*, 1996). Treatment means are stated to be significantly different only if the probability of differences occurring by chance were less than 5% ($p < 0.05$).

Results and discussion

Variety trial

At the sites Frankenhausen and Wulksfelde, comparable total yields of 35-36 t ha^{-1} were achieved (Table 3, mean of all varieties and both years). At the Lindhof site, a marginally lower yield level of 31 t ha^{-1} was reached. The two first early varieties Premiere and Velox, the medium-early varieties Agria and Marlen, and the late variety Marena, obtained very high yields on all three sites (Table 3).

Conspicuous differences between the varieties in terms of tuber size grades were evident (not shown). For processing of organically produced potatoes into French fries, a minimum tuber size of 35 mm is required, at which the fraction over 50 mm may not be less than 50% (Böhm, 2003). The tuber size-grade over 40 mm achieved in the mean of the varieties and years on all three sites was 73 to 79% of the total tuber yield. The requirement for the processing into French fries "50% of the tubers over 50 mm," was complied with only at the sites in northern Germany with the varieties Agria, Marena, Delikat, and on the Lindhof site additionally with Freya. In a comparison of the sites over the two years and all varieties, in Frankenhausen, the tuber fraction with a size over 50 mm accounted to only 20%. This was significantly lower than the 42% at the site Wulksfelde and 48% at the Lindhof.

Kolbe (2003) showed the correlation between weather conditions and starch content. So, high solar radiation, low temperatures or/and medium precipitation result in higher starch contents. The years 2003 and 2004 were characterised by entirely different weather conditions. The growing season 2003 was characterized by high solar radiation at low precipitation, while in 2004 lower temperatures prevailed with medium precipitation. Both weather situations resulted in a high concentration of starch in the tubers. The benchmark of 18% starch was exceeded by all varieties, particularly in 2003 (Table 4).

After harvest in both years, the concentration of reducing sugars fell below the required value of 300 mg 100 g^{-1} FM for processing into French fries, and also the varieties for processing

Table 3. Total tuber yield (t ha^{-1}) depending on varieties and the three sites Wulksfelde (WU), Lindhof (LI) und Frankenhausen (FH) in the years 2003 und 2004 (italic: crisps varieties).

		Total tuber yield (t ha-1)						
	Year	**1) 2003**			**2) 2004**			
Maturity Group	**Location Variety**	**1) WU**	**2) LI**	**3) FH**	**1) WU**	**2) LI**	**3) FH**	**Average Variety**
First early	Premiere	38.2	35.1	48.3	35.6	30.8	27.8	36.0
	Velox	41.7	32.6	45.5	35.0	28.4	35.9	36.5
Early	Camilla	34.4	29.0	40.9	29.6	25.8	27.5	31.2
	Carmona	*38.9*	*31.8*	*40.7*	*27.7*	*27.2*	*29.2*	*32.6*
	Delikat	32.6	27.0	36.7	26.6	29.3	31.6	30.6
	Diana	*32.1*	*31.6*	*39.3*	-	-	-	-
Medium early	Agria	43.6	41.1	42.9	36.6	26.8	32.4	37.2
	Freya	37.7	33.7	37.2	31.4	21.9	29.2	31.9
	Marlen	*43.9*	*38.1*	*41.5*	*37.8*	*26.2*	*30.6*	*36.3*
	Sempra	*32.4*	*30.0*	-	*29.5*	*26.3*	-	-
	Verdi	-	-	-	*27.9*	*25.4*	-	-
Medium late-late	Marena	43.6	38.3	43.2	31.7	24.4	33.1	35.7
	Saturna	*31.2*	*27.3*	*37.9*	*33.2*	*25.4*	*28.0*	*30.5*
	Sonate	35.3	33.5	-	31.2	27.0	28.6	-
S.E.		2.47						
Factor		**1**	**2**	**3**				
Location		35.6	30.0	36.0				
Year		37.8	29.9					

into crisps did not exceed the threshold value of 150 mg 100 g^{-1} FM (Table 4). After the 4-month storage period 2003-04, all varieties for processing to crisps showed a content of reducing sugars below the threshold and only the varieties Premiere and Delikat exceeded the benchmark for processing to French fries. After the storage period 2004-05, higher contents of reducing sugars were detected in the potatoes of the first early and early maturity group. The so-called 4°C-varieties (Putz, 2004), Sempra and Verdi, showed a low content of reducing sugars despite of the cold storage at 4°C. Data were comparable to the other varieties of the medium-early and medium-late maturity group. Principally, higher concentrations of reducing sugars have a negative effect on the quality like a brown discoloration, a bitter flavour of the end-product (Roe and Faulks, 1991), and an enhanced risk of high acrylamide levels. Therefore, the right choice of variety must ensure a compromise between yield and adequate quality of raw material.

Table 4. Content of starch and reducing sugars of the potato tubers to the time of harvest (H) and storage (ST) depending on varieties on the average of the three sites Wulksfelde (WU), Lindhof (LI) und Frankenhausen (FH) in the years 2003 und 2004 (italicized: crisps varieties).

		Starch (% of FM)				Reducing sugars ($mg\ 100g^{-1}$ FM)			
	Year	2003		2004		2003		2004	
Maturity group	Variety	H	ST	H	ST	H	ST	H	ST
First early	Premiere	19.9	21.2	20.5	21.5	74	404	167	740
	Velox	18.3	19.7	17.6	18.6	65	268	164	730
Early	Camilla	19.4	20.4	18.1	18.6	22	86	108	561
	Carmona	*19.8*	*21.4*	*19.0*	*20.0*	*22*	*150*	*72*	*434*
	Delikat	20.0	21.8	20.6	21.0	52	379	142	795
	Diana	*21.1*	*22.6*	-	-	*12*	*52*	-	-
Medium-early	Agria	19.5	19.9	18.6	18.8	13	34	19	141
	Freya	20.8	21.6	19.9	20.5	18	59	29	180
	Marlen	*20.7*	*21.1*	*19.8*	*20.7*	*17*	*35*	*21*	*137*
	Sempra[a]	*22.6*	*24.2*	*22.7*	*22.7*	*10*	*36*	*17*	*72*
	Verdi[a]	-	-	*22.8*	*23.3*	-	-	*37*	*134*
Medium-late-late	Marena	19.0	19.2	18.0	17.8	20	51	28	139
	Saturna	*21.7*	*22.3*	*21.7*	*22.2*	*16*	*55*	*35*	*156*
	Sonate[a]	21.5	23.0	21.5	22.2	20	87	27	253
S.E.[b]		0.64	0.63	0.64	0.63	15.4	54.8	15.4	54.8
Mean of varieties[b]		19.9	20.9	19.4	20.0	32	152	79	401

[a]only cultivated on the sites Wulksfelde and Lindhof;
[b]without the varieties Diana, Sempra, Verdi and Sonate

Previous crop management

The total tuber yield in 2003 averaged 37.5 t ha^{-1} and was higher compared to 2004 (29.1 t ha^{-1}). In 2003, there was a significant effect caused by the previous crops. The highest yield was achieved following the legumes field pea (41.4 t ha^{-1}) and alfalfa-ryegrass-clover (39.3 t ha^{-1}), which was significantly higher compared to the yield after the preceding crops oat (35.2 t ha^{-1}) or winter wheat (34.2 t ha^{-1}). In the year 2004 there were only small differences between the previous crops (Table 5). The contingent of the grading fractions > 40 mm and > 50 mm were higher in 2003 than in 2004. But there was no significant difference between the previous crops. The requirement for processing into French fries "50% of the tuber yield to be over 50 mm" was accomplished in 2003 for both varieties independent from the previous crop. In 2004 however, the percentage of the grading fraction did not reach the 50% benchmark.

Table 5. Total tuber yield and size grades depending on different crops preceding potatoes in the years 2003 und 2004.

Previous crop	Variety	Total tuber yield ($t\ ha^{-1}$)		Size-grade > 40mm (in %)		Size-grade > 50 mm (in %)	
		2003	2004	2003	2004	2003	2004
Field pea	Agria	43.9 a	29.9 ab	93.3 ab	88.1 a	66.9 a	43.1 a
	Marlen	38.8 bc	31.7 a	92.5 abc	73.5 c	56.7 b	44.2 b
Oat	Agria	37.4 cd	26.4 bc	91.4 b	85.7 ab	55.9 c	35.7 abc
	Marlen	33.1 e	25.4 c	91.9 b	79.5 bc	52.2 bc	33.0 c
Alfalfa-clover	Agria	41.0 b	30.1 a	93.4 ab	87.6 ab	64.4 a	39.2 abc
	Marlen	37.6 cd	29.0 ab	91.4 b	80.1 ab	56.6 b	30.9 c
Winter wheat	Agria	35.5 de	29.8 ab	90.6 c	86.1 ab	54.0 b	34.9 bc
	Marlen	33.0 e	30.2 a	94.2 a	82.4 ab	45.9 c	36.8 abc
Mean		37.5	29.1	92.3	82.9	56.6	37.2

ns = not significant; different characters indicate significant differences on a 0.05 level, tukey-test

The variety Agria showed a lower starch content (18.3%) than Marlen (21.0%). So the optimal range for processing Agria to French fries and Marlen to crisps was realised. Due to the dry weather conditions with high solar radiation, the starch content was higher in the year 2003 as compared to 2004 (Table 6). After storage, starch content increased on average of both years by 0.3%. The results documented that the influence of the year (weather conditions) and the choice of the variety is of higher relevance for the starch content in the potato tubers than the nitrogen supply which was affected by the previous crop.

Statistical analysis showed significant effects of previous crop, variety, year and storage on the content of reducing sugars in the potato tubers. The very low content at harvest 2003 and also after the 4-month storage period was probably caused by a low infestation with *Phytophthora infestans* connected with positive conditions for tuber maturation. In 2004 the early infestation with *P. infestans* disrupted the tuber maturation. Indeed, the content of reducing sugars was at a low level at harvest time, but increased during storage (Table 6).

Ryegrass-clover management

In 2003, the highest total yield was achieved by the mulch system according to the higher N-output of the ryegrass-clover residues (Table 7). This yield-advantage constituted up to 12% depending on the variety and management system. The variety Agria reached on average significantly higher yield levels than the variety Marlen (Table 7). This can be explained with the dry summer in 2003. Schittenhelm *et al.* (2006) distinguished two different potato types: a leaf type and a stem type. With incremental drought stress, the stem type (equivalent to Agria) wins an advantage in yield compared to the leaf type (equivalent to Marlen). No yield advantage of the stem type compared to the leaf type resulted from the relatively wet and cool growing season in 2004. According to the N-output of the ryegrass-clover residues, the management

Table 6. Content of starch and reducing sugars of the potato tubers depending on the crops preceding potatoes to the time of harvest (H) and storage (ST) in the years 2003 and 2004.

		Starch (% of FM)				Reducing sugars (mg 100g^{-1} FM)			
		2003		2004		2003		2004	
Previous crop	**Variety**	**H**	**ST**	**H**	**ST**	**H**	**ST**	**H**	**ST**
Field pea	Agria	19.0 a	19.2 a	17.5 b	16.9 a	15 ns	28 b	29 ns	135 d
	Marlen	21.7 c	22.0 cd	19.7 c	20.3 b	15 ns	15 a	32 ns	117 c
Oat	Agria	20.0 b	19.9 b	17.0 ab	17.5 a	14 ns	23 ab	26 ns	117 c
	Marlen	21.9 c	22.4 d	19.7 c	20.3 b	15 ns	16 a	24 ns	67 a
Alfalfa-ryegrass-clover	Agria	18.9 a	19.1 a	16.6 a	17.2 a	11 ns	41 c	29 ns	143 d
	Marlen	22.0 c	21.9 c	19.8 c	20.6 b	11 ns	21 ab	26 ns	108 bc
Winter wheat	Agria	19.8 b	20.0 b	17.0 ab	17.5 a	13 ns	22 ab	32 ns	147 d
	Marlen	22.0 c	22.3 c	19.5 c	20.0 b	13 ns	16 a	26 ns	102 b
Mean		20.7	20.9	18.4	18.8	13	23	28	117

ns = not significant; different characters indicate significant differences on a 0.05 level, tukey-test

system had no significant effect on the yield of the two varieties. Also, the grading fractions showed no statistical differences between the ryegrass-clover management systems. In both grading fractions, Agria had a higher percentage compared to Marlen. The requirement "50% of the potatoes have to be over 50 mm" was fulfilled in both years by Agria, except for the mixed system in 2003 (Table 7).

Potatoes suitable for French fries production should have a starch content between 14-18% starch in fresh matter and potatoes suitable for crisping even 16-20%, because that ensures high quality of the finished product, their crispness, texture and taste. The different weather conditions (2003: high solar radiation, 2004: low temperature with medium precipitation) in both years induced high concentrations of starch and dry matter. The required benchmark was exceeded by up to 3% (Table 8), that probably resulted in reductions of the quality score for the French fries and crisps (viscous texture). In 2003, a very low concentration of reducing sugars was achieved both after harvest and storage. After the 2004 storage, higher concentrations were detected, but only Marlen showed a significantly higher sugar content after cut mowing (Table 8).

Conclusion

The most important agronomic instrument for the organic farmer to secure high tuber yields and qualities is the right choice of variety. For processing into French fries the varieties Agria and Marena can be recommended for organic cultivation. They fulfil the specification of high

Table 7. Total tuber yield and size grades depending on different ryegrass-clover management systems as crop preceding potatoes in the years 2003 und 2004

Ryegrass-clover management	**Variety**	**Total tuber yield ($t\ ha^{-1}$)**		**Size-grade > 40mm (in %)**		**Size-grade > 50 mm (in %)**	
		2003	**2004**	**2003**	**2004**	**2003**	**2004**
3x cut mowing	Agria	36 a	32 ns	79 a	89 a	56 a	52 ns
	Marlen	32 ab	34 ns	73 b	87 ab	28 b	55 ns
cut mowing/mulching	Agria	35 a	30 ns	76 ab	87 ab	48 a	59 ns
	Marlen	31 b	29 ns	73 b	85 b	26 b	59 ns
3x mulching	Agria	38 a	30 ns	79 a	87 ab	55 a	58 ns
	Marlen	34 a	30 ns	74 b	85 b	33 b	51 ns
Mean		34	31	76	87	41	56

ns = not significant; different characters indicate significant differences on a 0.05 level, tukey-test

Table 8. Content of starch and reducing sugars of the potato tubers depending on different clover grass management systems as crop preceding potatoes at the time of harvest (H) and after storage (ST) in the years 2003 und 2004. FM = fresh matter.

		Starch (% of FM)				**Reducing sugars (mg 100 g^{-1} FM)**			
		2003		**2004**		**2003**		**2004**	
Ryegrass-clover management	**Variety**	**H**	**ST**	**H**	**ST**	**H**	**ST**	**H**	**ST**
3x cut mowing	Agria	19.9 a	19.2 a	18.8 b	18.8 b	13 b	27 ns	23 ab	124 a
	Marlen	23.2 b	23.7 b	21.7 c	21.9 d	11 a	25 ns	34 b	209 b
Cut mowing/mulching	Agria	20.0 a	20.2 c	17.9 a	18.4 ab	14 b	30 ns	28 b	135 a
	Marlen	23.1 b	23.9 b	21.3 c	21.4 cd	10 a	24 ns	27 b	107 a
3x mulching	Agria	19.6 a	19.1 a	18.1 ab	18.1 a	13 b	29 ns	12 a	112 a
	Marlen	22.8 b	23.1 b	20.8 c	21.2 c	8 a	21 ns	19 a	117 a
Mean		21.4	21.5	19.8	20.0	12	26	24	134

ns = not significant; different characters indicate significant differences on a 0.05 level, tukey-test tuber yields combined with a sufficient percentage of the size-grade over 50 mm and also high organoleptic quality profile, both at harvest and after storage. With regard to processing into crisps, cv. Marlen is highly suitable, as it was observed to have a good processing quality of raw material even after storage. Saturna as a late and a very common variety in the potato crisp industry did hardly accumulate reducing sugars but total and marketable tuber yield (> 40 mm) were on a low level. Depending on the degree of maturation, the level of reducing sugars in tubers exceeded the maximum tolerable concentration. In this connection, the so-called 4°C-varieties Sempra and Verdi are a good alternative to the common genotypes, although tuber yield of both varieties were not satisfying.

Furthermore, results confirm that every available agronomic measure to improve nutrient and water availability has to exploited in order to secure high yields at a satisfying quality level.

Leguminous crops such as field peas (*Pisum sativum* L.) with a following catch crop and ryegrass-clover leys can be recommended as preceding crops. Management of ryegrass-clover leys (cutting or mulching) does not have a significant impact on tuber yield, grading or quality. Agronomic measures did influence processing quality, yet the impact of the weather conditions on crop development was much more important.

In summary, it can be stated that organic potatoes for processing can be cultivated successfully when site- and farm-specific conditions are considered.

Acknowledgements

This project was funded by the "Bundesprogramm Ökologischer Landbau". Project-No.: 03OE003

References

Boehringer Biochemica (1995). Methoden der enzymatischen Bio-Analytik und Lebensmittelanalytik, Boehringer Mannheim.

Böhm, H. (2002). Effect of soil separation technique with integrated catch crops and manure on yield and quality of potatoes in organic farming. In: G. Wenzel and I. Wulfert (Eds.) Abstracts of papers and posters of the 15th Triennial Conference of the European Association for Potato Research (EAPR), Hamburg, 14 to 19 July 2002, 209.

Böhm, H., T. Haase and B. Putz (2002). Verarbeitungseignung und Ertrag von Kartoffeln aus Ökologischem Landbau (Processing suitability and yield of organically grown potatoes). Kartoffelbau 53(8): 323-329.

Böhm, H. (2003). Anbau von Kartoffeln zur industriellen Weiterverarbeitung (Production of potatoes for processing). In: K. Möller, H. Kolbe and H. Böhm (Eds.) Handbuch Ökologischer Kartoffelbau (Handbook Organic Potato Production). Österreichischer Agrarverlag: 158-164.

Kolbe (2003). Einfluss des Klimas (Influence of climate). In: K. Möller, H. Kolbe and H. Böhm (Eds.) Handbuch Ökologischer Kartoffelbau (Handbook Organic Potato Production). Österreichischer Agrarverlag: 14-19.

Littell, R. C., Milliken, G. A., Stroup, W. W. and Wolfinger, R. D. (1996). SAS system for mixed models. SAS Inst. Cary, NC.

Loges, R. (1998). Ertrag, Futterqualität, N_2-Fixierungsleitung und Vorfruchtwert von Rotklee- und Rotkleegrasbeständen. Kiel: Schriftenreihe des Instituts für Pflanzenbau und Pflanzenzüchtung der Christian-Albrechts-Universität zu Kiel, 262 p

Nitsch, A. (2003). Kartoffelbau (Potato Production). Agrimedia GmbH: 344 pp.

Putz, B. (2004). Kartoffelsorten für die Verarbeitung (Potato varieties for processing). Kartoffelbau 55(9/10): 358-364.

Roe, M.A. and R.M. Faulks (1991). Colour development in a model system during frying: Role of individual amino acids and sugars. Journal of Food Science 56: 1711-1713.

Schittenhelm, S., H. Sourell, F-J. Löpmeier (2006). Drought resistance to potato cultivars with contrasting canopy architecture. European Journal of Agronomy 24(3): 193-202.

Curriculum vitae – Herwart Böhm

Herwart Böhm studied Agronomy at the University of Giessen (Germany) from which he received a Diploma in 1984. After working on an organically-managed farm with potato production in northern Germany for two years, he went back to the University of Giessen and worked on a three-year project on minimal soil tillage systems *vs.* ploughing, with a focus on the effect of soil microbiology and the nitrogen cycle. He received a PhD degree from the University of Giessen (Germany) based upon his work on the microbiological assessment of soil-tillage-systems in different soil types. After he worked at the KTBL-Potato Research Station, he went to the University of Kiel, where he spent nine years researching organic potato production. He led a research program on seed potato production in organic farming. Other focal points of his research were the quality assessment of organically grown potatoes in terms to the choice of varieties, fertilizer treatments (manure and slurry) and the soil separation techniques. Field trials were also done with regard to the regulation of *Phytophthora infestans* and *Rhizoctonia solani* in organic potato production.

Herwart Böhm is co-author of the book "Handbuch Ökologischer Kartoffelbau" (Handbook Organic Potato Production) and has been a senior researcher at the Institute of Organic Farming at the Federal Agricultural Research Centre (FAL) since the year 2002. In a joint research project of the University of Kassel (Department of Organic Farming and Cropping), FAL (Institute of Organic Farming), University of Kiel (Institute of Crop Science and Breeding) and BFEL (Institute for Cereal, Potato and Starch Technology), different strategies for the cultivation of high quality potato raw stock for processing into French fries and crisps were examined.

Company profile – Federal Agricultural Research Centre: Institute of Organic Farming

The Institute of Organic Farming is one of 12 institutes of the Federal Agricultural Research Centre (FAL) in Germany. The Federal Research Centre, under the Federal Ministry of Food, Agriculture and Consumer Protection, pursues the goal of providing scientific basics to assist policy-makers in making decisions regarding nutritional, agricultural, and forestry policy as well as consumer policy, and extending scientific knowledge in these fields for the sake of public interest.

The Institute of Organic Farming was established in 2001 and is located in northern Germany between Hamburg and Lübeck. The animal herds of the 600 ha large experimental research farm of the institute include 100 cows with 50 cattle, 60 dairy goats with kids and 30 sows and fattening pigs.

Research topics of the Institute of Organic Farming are focused on organic husbandry, organic fodder production, impacts of environment depending on different land use strategies (conventional *vs.* organic farming), and socio-economic impacts. Objectives of research in plant and fodder production are the improvement of legume-based crop rotations in differently managed organic farm systems (without livestock, cow farm, pig farm) and the optimisation of the production of legume based fodder rations without any conventional fodder components. Research in organic potato production is focused on optimising yield and quality with regard to table potatoes and processing, as well as the development of strategies against the click beetles (*Agriotes* spp.).

Precision farming: Opportunities and limitations in potato farming

Klaus-Herbert Rolf
AGROCOM GmbH & Co. Agrarsysteme KG, Marketing/Sales, Potsdamer Str. 211, 33719 Bielefeld, Germany

Abstract

Precision farming - traditionally understood as site specific farming of agricultural land - is one of the most basic strategies used in modern agriculture, and over recent years a great variety of different possibilities have been developed in parallel with technical progress. Modern precision farming makes use of a complex range of technologies that can take into account the differing soil potential as well as optimising the quality and efficiency of farming.

Keywords: precision farming, (automatic) documentation, quality assurance, (mobile) data management, controlling systems

Introduction

Precision farming has passed through a strong development over recent years, mainly because it allows different interests and goals to be combined in one strategy. A farm may decide to embark on precision farming for legal, economic or sales reasons. One legal reason is that farms are continually facing tighter state controls in order to guarantee the retraceability of foods so as to protect the consumer and are also required to submit more and more documentation in connection with the state's commitment to environmental protection. These regulations leave very little room for scope as far as the nature and volume of the documentation is concerned and failure to comply can lead to considerable material and even existential sanctions. Precision farming instruments are generally based on automatic documentation solutions. And it is this element that makes it so interesting in connection with documentation requirements. Nevertheless, the main argument for precision farming is a financial one, namely to be able to look at data from various sources in the overall economic context (*cf*. Hasert, 2003). By taking into account site specific soil potential, costs and yields can be hugely optimised, for example through cost savings in the use of resources and machinery or through increased harvest yields. Efficient working processes and quality aspects can increase the earnings potential even more. The development of a quality management system using precision farming instruments gives many farms the chance to produce a quality brand. It is not just in the field of organic farming that this vast sales potential has been discovered. Figure 1 shows an overview of the instruments and technologies used for precision in agriculture.

Precision with a focus on potato farming

In many areas of crop husbandry, yield mapping is the first step towards site-specific observation of areas. However, its use in potato farming is not as widely developed as in grain farming, for

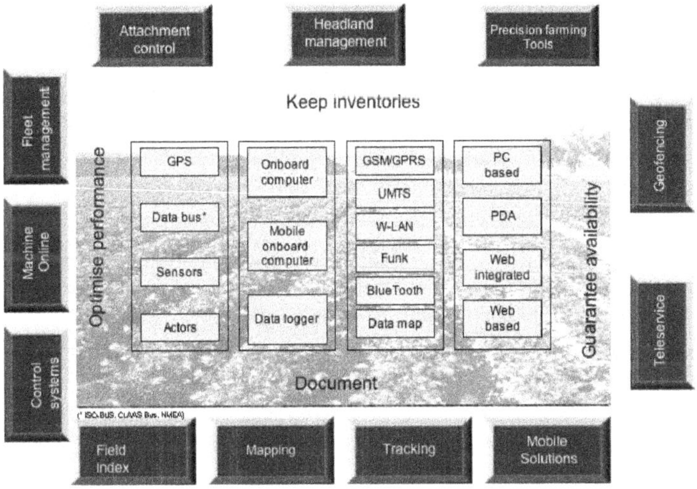

Figure 1. This chart presents an overview of the four main challenges faced by farms from the point of view of the practitioner. Farmers have a number of technical options at their disposal (shown in the centre of the chart) to help them to accomplish the tasks that derive from these challenges and to provide targeted support for precision in crop farming.

example, where the appropriate machine technology has been in use for years. Nevertheless, georeferenced soil samples can also be used to make assessments on differing harvest potential in potato farming, too. This can be taken into account during sowing, fertilising and crop protection. Each year the amount of information about the field also grows - not just in connection with crop rotation and the weather situation.

However, the fact that no sophisticated technical or automatic solutions exist that enable site-specific analysis of soil potential, does not means that precision farming is impossible for potato farmers. If we turn our attention to "precision agriculture", we can see three areas in which practitioners already successfully use precision farming instruments: firstly precise preparation and execution of work, secondly precise field cultivation and thirdly the integration of precise advice on crop cultivation.

Precise preparation and execution of work

As with most other areas of crop husbandry, it is common practice in potato farming to use a field index. The fast technical and highly functional development of field indices over recent years has led to quite revolutionary solutions that introduce a new level of quality into work processes. Field-related documentation of work done on fields is now carried out with the help of GPS technology, multifunctional on-board computers, generic standards (such as ASD, LBS+, ISOBUS, RS232, Bluetooth) and the integration of mobile handhelds. In other words: where once efficiency was restricted as a result of hand drawings or incompatible systems, data and machine systems now work together in perfect harmony. The modern farmer, or rather agricultural service provider, works with a multifunctional field index that combines mobile field management with GPS-aided navigation and surveying. Field-specific work schedules (for example on the use of fertilisers) are drawn up on the office computer, then transferred to a handheld that in turn transfers the data to the on-board computer of the attached implement and - after completion of the work - calls it up again. The driver automatically has control of all the documentation tasks on the field while actually on the job - including a precise record of cultivation measures, mapping, area surveying through to soil sampling and assessment. For example, it makes it easy to trace when and where work was done and what nozzle setting was used. A handheld system with a GPS aerial facilitates the entire work process: It automatically recognises the location of the file, shows the position of the driver, indicates when he enters and leaves the field (Figure 2). All the activities performed on the field, as well as breaks, are automatically documented. Individual field navigation assistants can be drawn up for temporary workers or new areas. These additional functions are a great advantage for farms that are constantly growing and where there is a fast crop rotations as is often the case with potato and vegetable farming. Apart from georeferenced documentation the new generation systems also provide important information and warning functions. Obstacles and nests of weeds are automatically displayed, so preventing delayed reactions. Water protection regulations are also accounted for: In fields next to streams and rivers the system automatically documents fertilisation in accordance with fertilisation and plant protection ordinances - including all the functions of a distance manager. It goes without saying that errors can be reduced to a minimum by using such a system, as well as saving time. In addition, the system makes logistics more transparent for many farms for the first time: journeys between fields and transport can be recorded reliably so that costs can be calculated on the basis of accurate data.

Precise field cultivation

In order to protect plants against pests, potatoes should only be planted on the same field at the most every four years. Crop rotations should include winter rape, winter wheat and winter barley, but also clover and sunflowers, depending on the type of soil and potato. When cultivating the soil in fields that have not been optically surveyed, GPS-aided parallel driving

Figure 2. The mobile GPS-aided application provides information and performs documentation tasks at the same time.

systems have proven to be extremely efficient instruments in precision farming. Depending on the use, area and size of the farm, practitioners rely on tractor steering aids, automatic steering systems or systems with additional base stations that work with a level of accuracy of up to 5 centimetres, for example when planting potatoes (Figure 3). The background: Overlapping tracks waste valuable time and money (Table 1). Parallel driving systems counteract this very effectively. Thanks to GPS the position of the tractors on the field can be located very exactly and it can be guided along a specific line - either in a straight line or along curves.

Optical steering systems are very useful. In potato farming they can be used to apply fertilisers and pesticides. With the help of a camera the driver is carefully guided between the ridges with a tolerance of just a few centimetres. The vehicle is automatically guided parallel to the drill track as the system recognises the track tracer (Table 2).

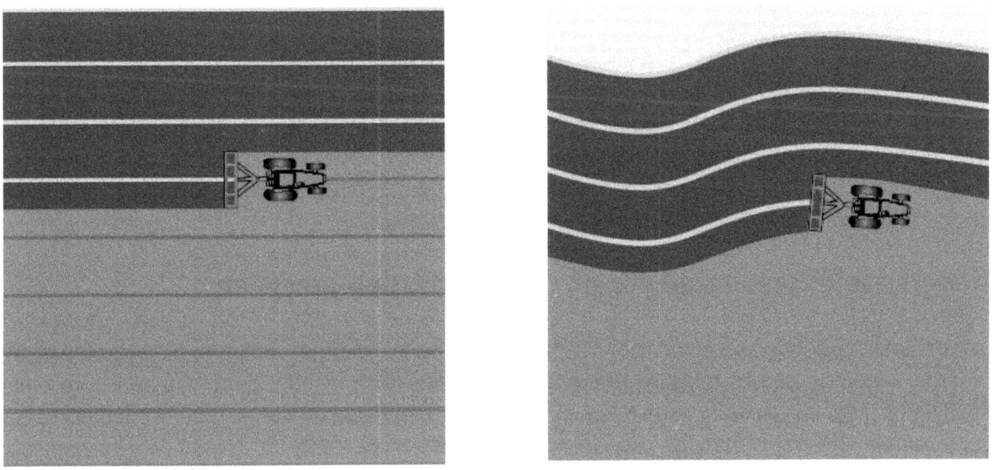

Figure 3. Before starting work the driver selects a direct A to B route or contour route.

Table 1. Savings from parallel driving systems for different cultures. Assumptions: field size 5 ha, no cultivation, overlapping up to 10 m working width 4%, over 10 m working width 7%, tramlines 3.4%. (Hüter et al. *(2005).*

Type of culture	Saving
Winter wheat	17.77 €/ha
Winter wheat with slurry	19.34 €/ha
Winter barley	16.51 €/ha
Greenland	19.87 €/ha
Sugar beet	7.02 €/ha
Grain maize	1.18 €/ha

Table 2. Economic effect of damage reduction and time saving at different crops using a photographic sensor (Kuhlmann 2005).

Culture	Economic effect of damage reduction (€/ha)	Economic effect of time saving (€/ha)	Total effect of application (€/ha)
Maize	1.32 - 6.58	3.47	4.79 - 10.05
Sugar beet	3.66 - 18.32	4.77	8.43 - 23.09
Potatoes	0.87 - 4.35	3.90	4.77 - 8.25
Lettuce	5.25 - 26.25	9.10	14.15 - 35.35
White cabbage	10.72 - 53.62	18.35	29.07 - 71.97
Strawberries	16.94 - 84.72	Ca. 12.-	28.94 - 96.72

The advantage of all steering systems is that even less competent drivers can work very efficiently and can devote their full attention to the tool settings. Steering systems can also be used for mechanical weed control, the preferred method of control for organic farmers. This was previously done by hand and was very time-consuming. With the additional aid of sensors, crop measurement and crop-specific use of fertilisers and pesticides can in future be combined in one task, as is already common practice in grain cultivation.

Integration of precise advice on crop cultivation precision

Server, web and telecommunication technologies have become common in agriculture in order to speed up and to improve the quality of information management (Figure 4). In crop husbandry they are primarily used to provide ongoing individual advice in daily work processes. For example, the farm manager and consultant can access centrally stored information on specific fields at the same time and no matter where they are. The consultant can make his recommendations directly based on the specifications of the field, for example recommendations about fertilisers or pesticides that are not recommended for all early potatoes but which achieve the best results for the CE type on this location with a high sand content. Compatible Software components allow the farmer to record the work directly in his field index. All he has to do is to confirm and the documentation is finished. This software compatibility represents a new milestone in inter-farm cooperation. This strict system of data management can be combined in a processing chain that integrates the individual farm, cooperatives and buyers. Time and money can be saved at the same time as fulfilling official documentation requirements.

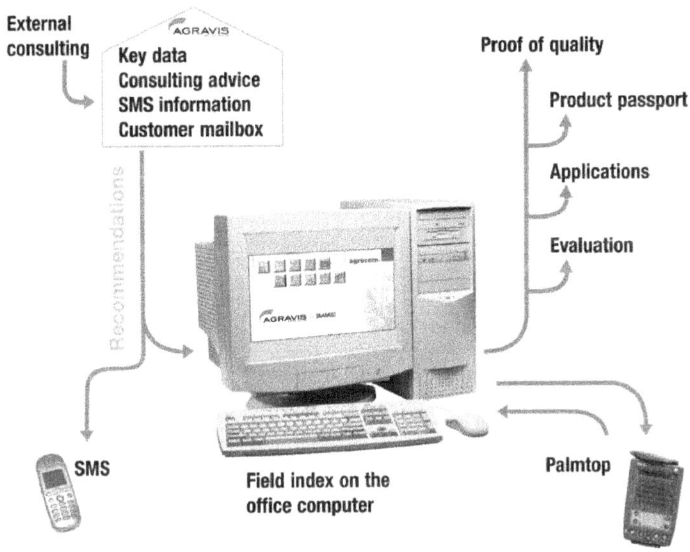

Figure 4. Application model from the web-supported "Raiffeisen Ackerbau Management Informationssystem RAMIS" that combines the basic functionalities of a field index with the latest professional cultivation recommendations.

Conclusion

The goal in potato farming is to implement instruments that enable precise preparation and execution of work, precise field cultivation and the integration of precise advice on crop cultivation.

References

Hasert, G. et al. (2003): «Zukunftsträchtiger Ackerbau, Systeme der computer- und GPS-gestützten teilflächenspezifischen Bewirtschaftung praxisnah bewertet» (Pioneering husbandry, a practical evaluation of computer and GPS-aided systems for site specific land cultivation). Deutscher Bauernverlag, Berlin

Hüter, J. et al. (2005): «Elektronik, Satelliten und Co, Precision Farming in der Praxis» (Electronics, Satellites and Co, Precision Farming in practice) KTBL publication

Kuhlmann, Bernd, Dipl.-Ing. agr. (2005): „Ökonomische Beurteilung und strategische Überlegung zum Einsatzpotential des Bildsensors" (Economic evaluation and strategic considerations on the savings potential of photographic sensors), Institute of business administration in agriculture and nutritional science at Justus-Liebig-University Giessen

Curriculum vitae – Klaus-Herbert Rolf

Klaus-Herbert Rolf grew up on a farm in Westphalia and, after completing an apprenticeship in agriculture, studied at the Faculty of Agriculture of the University of Paderborn in Soest. After graduating in agriculture in 1986, he joined the marketing division of Klöpper and Wiege in Lemgo, a company specialised in the development and marketing of agricultural software. Since 1989 he has been in charge of marketing. He has been with the Bielefeld company AGROCOM GmbH & Co. Agrarsystem KG, a 100% subsidiary of CLAAS company since its foundation in 1998, where he is also responsible for marketing.

Company profile – AGROCOM

AGROCOM can look back on many years of experience in the field of systems technology for agriculture. The company provides software solutions for crop husbandry and animal production, ranging from start-up programmes through to all-round solutions, from networks to internet-based databases. Hardware and machine components are also part of the company's range of services. Farming made a great leap forward when computer technology was introduced into agricultural processes and now there is a great demand for complex networks as part of computerised farming. The goals of the industry are safety and efficiency, based on complete and unbroken documentation. In view of the necessity to optimise unit costs and also the public demand for retraceability, it is essential throughout all stages of production to collect and process information, to link this information and to optimise the quality of presentation - if possible fully automatically. AGROCOM helps farms to meet this challenge with sophisticated products that support cooperation between farms and manufacturers, both in the domestic economy and between the various stages of production. All the products and systems are designed to meet the demands both of beginners and professionals; they can be extended and grow with the farm.

Crop protection and fertilisation

Relevant diseases in potato production

Frank Niepold
Federal Biological Research Centre for Agriculture and Forestry (BBA), Institute for Plant Protection in Field Crops and Grassland, Messeweg 11-12, D-38104 Braunschweig, Germany

Abstract

Symptoms of major virus, bacterial and fungal diseases infecting potatoes are demonstrated. From the group of potato viruses potato virus Y (PVY) and potato leaf roll virus (PLRV) are presented. The PVY strain ntn causes severe necrotic symptoms on tubers, PLRV infects the whole plant. Both viruses are transmitted by aphids. The soft rot disease of the tubers and black leg is caused by the bacterium *Erwinia carotovora*. The quarantine disease caused by the bacterium *Ralstonia solanacearum* can be detected during early infection by a loss of turgor and by wilting of the whole plant at a late stage, since the bacterium is a vascular pathogen. There are many ways of spread of this bacterium that poses a permanent threat of infection. Another potato quarantine bacterium is *Clavibacter michiganensis* ssp. *sepedonicus*, the causal agent of the potato ring rot disease. This bacterium grows optimal at low temperatures and it spreads via xylem throughout all parts of the plant including daughter tubers. Potato wart caused by *Synchytrium endobioticum* is a quarantine disease as well The quarantine status of *Synchytrium endobioticum* is due to the so-called winter sori which survive and remain infective in the soil for 20 years or more. No appropriate chemical treatments are available, only breeding for resistance. The dry rot disease is caused by the fungus *Fusarium* infecting mainly stored tubers. The fungus which causes the most severe damage on potatoes worldwide is *Phytophthora infestans*, the causal agent of late blight disease. There are two mating types, increasing the genetic variability and the survival (oospores) of this fungus until the next growing season. Tubers are also infected and are a latent reservoir for further infections during the next growing season. Early blight is caused by *Alternaria* species and the symptoms are characteristical brown spots with concentric rings on the leaves. *Rhizoctonia solani* is the causal agent of black scurf. Predominant black sclerotia are found on the skin of infected tubers. *Rhizoctonia solani* can overwinter in the soil and therefore can only be partly controlled by chemically dressing seed tubers. To prevent a further spread of the disease in storage, a fast drying and cooling of the tubers is recommended. Silver scurf (*Helminthosporium solani*) lately has become an increasing problem during storage.

Keywords: symptoms, storage, potato virus Y, potato leaf roll virus, *Erwinia carotovora*, *Ralstonia solanacearum*, *Clavibacter michiganensis ssp. sepedonicus*, *Synchytrium endobioticum*, *Fusarium* sp., *Phytophthora infestans*, *Alternaria* sp., *Rhizoctonia solani*, *Helminthosporium solani*

Introduction

When the potato plant was imported from South America in the second half of the 16th century, its diseases followed later. The appearance of the late blight disease in the years 1845/1846 in Ireland led to a drastic decline in yield, and subsequently to a famine that made 3 million Irish people suffer.

In order to have an overview of the potato's major diseases, pathogenic micro-organisms, *i.e.* viruses, bacteria and fungi are presented. Since the potato is very demanding with regard to plant protective measures, knowing its diseases is of importance and may subsequently be of great interest for growing and storing potatoes.

Viruses

One of the most important virus diseases of potatoes is the potato virus Y (PVY) which can cause severe losses. This virus is known to have three different strains, but the most severe damage is caused by strain ntn (new tuber necrosis). This specific virus is transmitted by many different aphids. It attaches to the stylets of the aphids, directly transmittable or in some cases able to remain viable for about 24 hours. The virus can also survive in tomatoes and other *Solanaceae* plants. The only possible control of this disease is to interrupt the transmission by preventing aphids from settling on potatoes or to grow virus-resistant potato varieties.

Another virus disease is the potato leaf roll virus (PLRV). This virus is persistent, and is therefore only transmittable by aphids, sucking for at least 1 to 2 days on PLRV-infested plants. At the first stage of infection the virus causes the upper, youngest potato leaves to roll up. At a later stage of the infection the leaves are completely rolled up, and the plant starts yellowing. When seed potatoes are infected by this virus, a yield loss of 20% may occur. Here again, the only way to control this virus is to control the transmitting aphids or to grow virus-resistant potato varieties.

Bacteria

An important bacterial disease is the soft rot disease of tubers caused by *Erwinia carotovora*. This bacterium can also infect potato plants in the field, and the disease then caused by *Erwinia carotovora* is called black leg. Here, the stems can easily be pulled out because they are macerated at the stem bottom by the bacterium. At a later stage the whole plants are wilting. This bacterium is usually described as a wound parasite and does preferentially infect injured tubers. The disease starts with brown dents and, at optimal temperature conditions, can proceed to macerate the whole tuber. Then the infected tuber still has an intact skin. A soft, paste-like mass will protrude when squeezing the infected tuber. In association with this pathogen other bacteria or fungi are usually occurring, causing an unpleasant smell and sometimes forms slimy exudate. The bacterium *Erwinia carotovora* can survive latently inside or on the surface of seed potatoes. During the vegetation period, the bacterium can be transported to new-grown daughter tubers and infect them. Usually the way of preventing *Erwinia carotovora* infections in storage is to dry and cool the tubers allowing the production of cork (suberin) which then entraps the bacteria. However, to protect tubers from *Erwinia carotovora* infections they should be allowed to mature in the soil, forming a hard skin. Therefore, a delayed harvest of the tubers is recommended because this bacterium is one of the major storage diseases. Generally, no antibiotics are allowed to control any plant pathogenic bacterium. probably, at least in the lab it worked whit a number of antibiotics.

The symptoms produced by this bacterium should not be confounded with the symptoms of another disease, the so-called bacterial brown rot disease caused by the quarantine bacterium *Ralstonia solanacearum*. Since this bacterium preferentially grows in the vascular rings it blocks the water and nutrition supply, resulting in the characteristic wilting of the foliage on the fields. An early stage of infection can be recognised by a loss of turgor at higher temperatures at noon which disappear when the temperature decreases. At a later stage of infection the potato plant cannot recover any more and will wilt yellow or finally brown. This bacterium preferentially infects through wounds and can grow latently, *i.e.* without any symptoms in tubers and can even survive in the soil. The bacterium is also transmittable by irrigation. Infected tubers and stems can be diagnosed by cutting and squeezing the cut surface. Then the bacteria extrude a grey or brown slimy ooze. Infected tubers do not sprout, and many other plants (tobacco, weeds, sugar beets) can serve as host , presenting a permanent infection threat for potatoes.

The only possibility to control this disease is to use certified seed potatoes and to accept the quarantine measurements imposed on contaminated fields. To prevent a further spread of the bacterium, all instruments and containers which have been in contact with the contaminated lot should be disinfected. Anyhow clean and prudent working is a requirement of good agricultural practice to avoid spreading of any potato pathogenic micro-organisms.

Similar requirements are necessary for another potato quarantine bacterium, *Clavibacter michiganensis* ssp. *sepedonicus*, the causal agent of the potato ring rot disease. This disease is named after the dark ring appearing in diseased potato tubers, representing the infected vascular bundle. The bacterium needs a temperature of 19° to 23°C for optimal growth, and it spreads via xylem to all parts of the plant including daughter tubers. The detection of ring rot-diseased plants in the fields can be cumbersome since the symptom is again a slow wilting of usually a single stem of a potato plant. Like *Ralstonia* this bacterium can only be controlled by strict quarantine measurements to prevent contamination of not infected tubers, and by using certified seed potatoes. Both quarantine bacteria can be detected routinely by immune-fluorescence techniques and the polymerase chain reaction (PCR).

Fungi

Another quarantine organism of potato is a fungal disease named potato wart which recently emerged to become an increasing problem for potato growers. Potato wart is caused by a fungus *Synchytrium endobioticum* generating typical galls on infected tubers. The fungus does not produce a mycelium but forms zoospores which infect only young sprouts. Infected tubers form galls instead of stems containing infective organs, the so called sori. Parts of these sori are the winter sori which can survive in the soil and remain infective for up to 20 years which is the reason for its quarantine status.

There are no appropriate chemical treatments applicable to control this disease. However, there are resistant potato varieties available which unfortunately do not have a large market volume. Therefore, the only way to control this disease is a breeding program for varieties resistant to *Synchytrium endobioticum*.

In potato storage, infection can occur by another fungal pathogen, *Fusarium* species. About 30 species worldwide can cause the so-called dry rot disease. Usually, its symptoms are only visible at later stages during warm storage conditions. The first symptoms of this disease show a darkening of the infected areas. Then a subsequent shrinking of the infected tubers can be observed, caused by the fungal uniform growth throughout the whole tuber which subsequently results in loss of water. Because of this water loss the infected areas turn dry, and a typical white-blue or white-yellow mycelium is formed. *Fusarium* species enter the tubers via injuries since they are typical wound parasites. The fungus forms spores that persist in the soil and can survive there for several years, still being infective. In addition, after harvest, the attached soil can be contaminated with spores of this fungus and can lead to infection during storage. After storage of seed potatoes this fungus can be transferred to soils because spores are attached to the skin of the tubers. A *Fusarium*-infection can prevent tubers from sprouting, or in some cases even prevent growth.

To avoid the *Fusarium* infections a careful harvest and handling of the tubers is recommended to reduce injuries. In addition, tools and storage rooms should always be cleaned before storing tubers from a new harvest.

One of the most important fungal diseases of all potato growing areas is the late blight disease, caused by the oomycete *Phytophthora infestans*. This fungus is destructive, and a great amount of fungicides are applied worldwide to control this disease. The symptoms are typical brown lesions, and during early morning hours a kind of mildew mycelium can be found at the lower side of the leaves or at parts of the stems. The fungus produces sporangia which contain zoospores. Both infective organs need water to spread. At low morning temperatures the zoospores are transported in small droplets by rain or irrigation over distances of up to several kilometres on windy days. Usually, an infection starts in the field from latently infected volunteer plants or from latently infected plants growing on dumps *etc.* Usually, latently infected potato plants show no visible symptoms but *Phytophthora infestans* is already growing inside. It seems that the so-called "stem-*Phytophthora*" is the first symptom of latently infected tubers.

The sexual formation of oospores, the product of a mating between two different mycelia types (A1 and A2), is the reason for the survival of *Phytophthora infestans* for an extended period of time in the soil, since the cell walls of these oospores are durable. Therefore, a subsequent infection can take place from oospore-infested soils over a longer period of time. In addition the genetic variability increases by recombination that occurrs during mating. *Phytophthora infestans* is also able to infect tubers to produce brown rotting tubers. Due to technical advances in storage, latently infected potato tubers can hardly form symptoms but can survive and subsequently be planted as supposedly symptomless in the next season, but are a source of the stem-*Phytophtora* in the new vegetation period.

An additional source of infection can arise from infected tomatoes since *Phytophthora infestans* can infect both plant species. Therefore, mutual infections can occur in areas where both crops are grown.

Besides chemical control measurements, there are only moderate possibilities to reduce this disease by planting resistant potato varieties.

Early blight disease, caused by the fungus *Alternaria* species, recently is of increasing importance.. Its symptoms usually appear on the lower leaves of the potato plant from where the fungus spreads to the upper leaves. The symptoms are distinct and can be diagnosed very easily. Large brown spots with concentric rings are characteristic for this disease. Since the rest of the healthy leaves continues to grow, these spots may fall out and infect tubers via its conidia. Massive infections will kill the plant while the conidia of *Alternaria* survive in dead leaves or in stems. Young tubers are easier infected by this fungus than older ones. Infected tubers can be diagnosed by the formation of hard tuber surfaces. Here again, beside chemical treatment a longer maturation of the tubers in the soil by allowing to form a firmer skin can prevent *Alternaria* infection.

Black scurf or *Rhizoctonia* cancer is caused by the fungus *Rhizoctonia solani*. Predominant black sclerotia are formed on the skin of infected tubers which can not be easily removed. This fungus sometimes causes malformations on tubers that are visible after harvest. After sprouting the stems will be infected in the ground from the sclerotia present on the mother tubers, resulting in a delayed emergence or dying of all sprouts. At a later stage the symptoms of this disease are a stunting or roset formation of the remaining plants, or a production of aerial tubers on the stems. The infection takes place in cool, moist soils since *Rhizoctonia solani* can overwinter in the soil. Black scurf only partly can be controlled by chemical dressing of the tubers. To prevent a further spread of the disease in storage, a fast drying and cooling (3°C) is recommended.

Silver scurf (*Helminthosporium solani*) lately has become an increasing problem in the storage of potato tubers. The fungus infects the tips of the stolons in the soil and can prevent plant growth. Silver scurf is hardly visible on tubers at harvest, since the temperature of the soil is usually too low for symptoms to develop. However, when the tubers are in storage, silver scurf symptoms can develop, resulting in large silver grey spots. Since the fungus needs moist and low temperature conditions for its development, these conditions are present when the tubers are cooled down. Silver scurf infection causes a loss of water of up to 5% (NOT MORE?), resulting in a quality problem because of a shrinking of the tubers. Here, similarly as for black scurf, fast drying and cooling (3°C) of the tubers is recommended.

Beside the fungal diseases described here there are also other fungal potato diseases such as *Colletotrichum*, *Pythium*, *Verticillium* and *Phoma* which locally can cause severe damage. In addition to diseases, there are pests such as nematodes (cyst and root-knot nematodes) and insects (aphids, moths, Colorado beetle) which severely can damage potato. The potato plant is very attractive for pathogens, and therefore demands various protection measurements during growth in the field as well as during storage.

Conclusion

Potatoes are attacked by numerous plant pathogens specialised towards this plant. Knowing the symptoms caused by the pathogens makes it easier to diagnose the pathogen which has infected the potato. Control measurements can then be easier performed. Simple measurements like keeping the tubers in the soil as long as possible before storing them can overcome some of the disease problems caused by bacteria and fungi.

References

Farbatlas Krankheiten und Schädlinge an landwirtschaftlichen Kulturpflanzen, (1999), B. Schöber-Butin, V. Garbe, G. Bartels, Eugen Ulmer Verlag, Stuttgart, 240 pp.

Krankheiten und Schädlinge der Kartoffel, (1990), W. Radtke and W. Rieckmann, Th. Mann Verlag, Gelsenkirchen-Buer, 168 pp.

Compendium of Potato Diseases, (1981), W. J. Hooker, American Phytopathological Society, St. Paul, Minnesota, USA, 124 pp.

Curriculum vitae – Dr. F. Niepold

Frank Niepold is married with two children and was born on 04 November 1952 in Hannover, Germany. From 1959 to 1972 he received his elementary and high school education, after which he served in the army from 1972 - 1974.

He studied Agriculture from 1974 until his diploma in 1978 at University Göttingen. After his graduation, he continued with a PhD study at the University Göttingen in the field of bacterial plant pathology, that he finished in 1981. He received a post doctoral stipend at the Oregon State University, Corvallis, Oregon, USA, where he did research in the field of host-parasite interaction and characterisation of pathogenicity genes from 1982 - 1985. In 1986 he went back to Germany to study serological detection methods for plant pathogenic bacteria at the Technical University Munich.

Since 1988 he works at the Federal Biological Research Centre of Agriculture and Forestry, where he does research on the release of genetically engineered micro-organisms, development of detection methods by using PCR techniques. His present work is diagnosis of potato pathogenic bacteria and fungi, especially of *Phytophthora infestans* and *Synchytrium endobioticum*.

Company profile – Federal Biological Research Centre for Agriculture and Forestry

The Federal Biological Research Centre for Agriculture and Forestry is both an independent higher Federal authority and Federal Research Institution. It is subordinate to the Federal Ministry of Food, Agriculture and Consumer Protection (BMELV). Its tasks are stipulated in the German Crop Protection Act and the Gene Technology Act. The BBA's research in the vast field of plant protection and plant health contributes to sustainable maintenance of the cultivated landscape and people's quality of living.

Regulatory Activities

In the framework of plant protection product authorization, the BBA evaluates products with regard to their efficacy, phytotoxicity, resistance management and actual benefit. Its agreement is required for any product authorization. The BBA also tests plant protection equipment and keeps an official list of tested plant protection equipment. It develops methods to test crop varieties for their potential to resist diseases and supports the general testing of plant varieties.

The BBA is also involved in the approval procedure for releasing and marketing genetically modified organisms. Another important field of work includes regulatory activities for plant inspection and plant quarantine within the European Union (EU). The BBA's administrative tasks are closely connected with research activities, providing a solid basis of up-to-date scientific and technological know-how for its decisions. An important field of work for the BBA is the improvement of integrated plant protection concepts under which biological, cultural and other techniques would help to reduce the use of chemical products to a necessary minimum. The aim is to secure the production of healthy plant products.

BBA scientists develop computer-aided forecasting models to estimate risks of pest attacks and damages, and models to facilitate the optimum use of plant protection products. The BBA research is also concerned with preservation strategies for healthy plants in organic farming, urban environments and forests. Plant protection begins with diagnosing the causes of damage occurring in plants. BBA scientists investigate the functions of pest organisms and diagnose plant diseases. They work on suitable biochemical, molecular-biological and serological methods to detect important pests and diseases.

In the context of the release of genetically modified organisms, the BBA conducts comprehensive research on biological and ecological safety. BBA scientists study the effects and fate of plant protection products in the cultivated landscape and work on post-registration monitoring methods of plant protection products. With its research the BBA contributes to safeguarding a diverse cultural landscape and biological diversity.

Research

According to the political foci and the research plan set out by the Federal Ministry of Food, Agriculture and Consumer Protection (BMELV), the BBA's research work is aimed at developing methods and techniques serving to protect plants and plant products and avert any risks to the health of humans, animals and the natural balance which might arise from plant protection measures.

Early blight: Effect of different parameters on the disease development

H. Hausladen and J. Leiminger
Chair for Phytopathology, Life Science Center Weihenstephan, Technical University Munich, Am Hochanger 2a, 85350 Weihenstephan, Germany

Abstract

Early blight is a major foliar disease of potatoes, causing premature defoliation. Various factors have an influence on the development of early blight. In this work, effects of various parameters like plant age as well as fungicide concentration on the disease progress are shown. Two-year field trials were carried out in order to determine the influence of different plant ages on disease progress. Plant age of potatoes was manipulated by late planting. A strong correlation was found between maturity and disease severity due to early blight. Earlier planted potatoes, which matured earlier were more susceptible, and later maturing potatoes were more resistant. The most effective parameter influencing the disease progress is the use of fungicides. In particular there is a need to develop effective fungicide strategies in order to control this disease. Increased early blight severity, despite the use of protectant fungicides which contain mancozeb, prompted us to study the influence of different dose rates of mancozeb on the development of early blight on potato foliage. In field trials the efficiency of six different dose rates of mancozeb was studied. The dose rate of 1500 g mancozeb per hectare was seen as a reference. Significant differences in disease severity among different fungicide dose rates were observed by the end of the season. There is a strong correlation between dose rate of mancozeb and early blight control. Mancozeb at lower concentration was less effective than higher concentrations.

Keywords: early blight, *Alternaria*, chemical control, dose rate, plant age

Introduction

For agriculture potato is still an important crop. In 2005 over 276,000 hectares of potatoes were cultivated on German agricultural land (2.3% of total German arable farm land). Despite a declining cultivation, potato is the second most important staple food in Germany after grain. Apart from the widespread potato disease late blight, early blight has caused increasing problems for German farmers in the last years. Since the last six years a monitoring programme has been accomplished in potatoes throughout Germany, which documents the occurrence of early blight in all German potato growing areas. Within the last years an increase in disease frequency was observed and the disease was established as a relevant and destructive pathogen (Hausladen *et al.*, 2004). Early blight became a more and more important disease in most of the potato growing areas.

Early blight of potatoes, caused by the fungi *Alternaria solani* and *Alternaria alternata*, results in significant yield losses. Different factors, depending upon environment and plant physiology, have a severe influence on the disease progress of *Alternaria*.

The disease is still primarily managed by the application of foliar fungicides. Therefore, a fungicide spray programme should be used from early in the growing season to vine kill (van der Waals *et al.*, 2001). An early blight research project at the Technical University in Munich-Weihenstephan deals with the basics of the epidemics as well as fungicide strategies in disease control. Field experiments were carried out in 2004 and 2005 to evaluate the efficacy of different fungicide concentrations and the influence of plant age on early blight.

Disease cycle

The fungus survives the winter as conidia or mycelium on infected plant debris in soil or on seed (Rotem 1994). At favourable conditions spores are formed and are either wind blown or splashed onto plant surfaces where infection occurs at first on lower leaves. On potatoes *Alternaria* causes big brown and necrotic lesions, clearly angular and bounded by the nerves (Figure 1). Initial infections are most frequent on older leaflets. Lesions begin as dark brown to black spots, about 1 mm in diameter. The spots develop in a somewhat irregular shape and usually show concentric rings ("target spot") (Waals, van der *et al.*, 2004). As disease progresses, infected leaflets may turn yellow and either dry up or fall off. In some years stem infections can occur with symptoms similar to those on the leaves.

Figure 1. Early blight symptoms

Alternaria solani belongs to the large-spored group within the genus, and produces simple, single-borne conidia that contain long and occasionally branched beaks (Ellis and Gibson 1975) (Figure 2). By contrast, *Alternaria alternata* is one of the smaller-spored species, whose conidia are multiseptate, darkly pigmented and found as short chains borne on conidiophores, rather than single conidia (Petrunak and Christ 1992) (Figure 3).

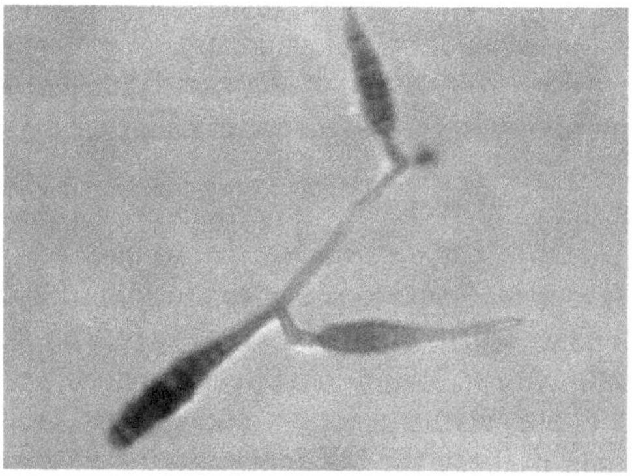

Figure 2. Single borne conidia of Alternaria solani

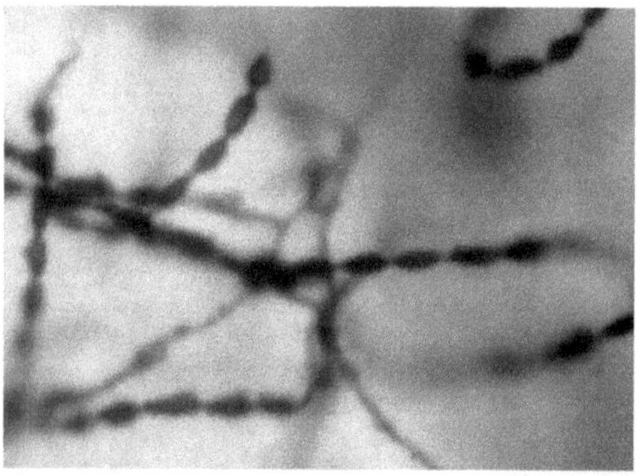

Figure 3. Conidia of Alternaria alternata *formed in chains*

Recognized primarily as a foliage pathogen, epidemics occur when the weather is warm and dry with short periods of high moisture (frequent rain and abundant dew) (Venette and Harrison 1973, Chelkowski and Viskonti 1992). Only by regarding symptoms it is not possible to distinguish if the necrotic spot is caused by *Alternaria solani* or *Alternaria alternata* (Hausladen 2004). Morphological analysis of the formed conidia or PCR-technique are required for a clear differentiation of both pathotypes. For this reason in field trials early blight is imputed to *Alternaria ssp.*.

In their requirements concerning growth conditions both pathotypes are very similar. Already three to four weeks after crop emergence first symptoms can be found in the field on older leaves in the lower leaf levels of the canopy (Johanson and Thurston 1990). Under favourable weather conditions disease progresses and lesions are also formed on leaves at higher leaf levels. If blight is severe, whole leaves can be infected and plants may be entirely defoliated. Heavy infections can cause yield losses over 30% (Johnson *et al.*, 1986). Especially in late maturing potato varieties early blight can reduce starch content through premature defoliation. Because *Alternaria* produces toxins that diffuse into host tissues, enlarging lesions are often surrounded by a narrow chlorotic halo (Rotem 1994).

The most effective disease control has been the frequent application of protectant fungicides. Therefore, applications have to start at an early stage of the disease (Waals, van der 2001). At this time early blight is still inconspicuously in the occurrence of symptoms and is found only on the lowest leaves. Control of this polycyclic disease depends primarily on multiple fungicide applications.

Materials and methods

Two field trials were carried out in 2004 and 2005 to evaluate the efficacy of different fungicide concentrations as well as the influence of plant age on early blight. Trials were designed as a randomized complete block. Each plot reached over 8 rows in width and 8 m in length. Fungicide treatments with mancozeb were replicated 3 times whereas the trials with different planting dates were replicated four times. The experiment was fertilized, cultivated and managed according to general agricultural practices. In order to prevent late blight the fungicide RanMan (400g cyazofamid/liter) was applicated as a coverspray every 8 to 10 days at a dose of 0.2 liters per acre. Disease progress of early blight was not affected by the use of RanMan.

In order to investigate the influence of plant age on the progress of early blight, potatoes were planted at two different dates with approximately four weeks difference. In 2004 potatoes were planted on March 30th and on May 2nd. Crop emergence differed in about two weeks. In 2005 potatoes were planted on April 3rd as well as on May 2nd. Here the first layer emerged on May 16th whilst the later ones emerged on May 28th. Accordingly the later planted potatoes were 12 days later in its growth in comparison to the earlier planted potatoes.

Disease progress was observed from beginning of early blight symptoms until the end of the season. Both centre rows of each plot were harvested and yield as well as starch content were assessed. Fungicide trials included an untreated check (only coverspray), the active ingredient mancozeb was applied at different rates of 500, 750, 1000, 1500, 2000 and 4000 gram per hectare and application. The fungicide treatments were initially applied when first early blight symptoms appeared on potato leaves and continued every two weeks for the remaining growing season. Fungicides (400 liters of water/ha) were applied with a portable backpack-sprayer.

Early blight was allowed to develop naturally during the course of the growing season. Disease progress was observed by weekly ratings with appearance of the first disease symptoms. Single plants of the central plot rows were rated for the percentage of the leaf area covered by early blight lesions. All disease rating dates and disease severity values were used to calculate disease progress and the area under the disease progress curve (AUDPC).

Results

Effects of fungicide concentration on disease severity of early blight

The efficacy of various fungicide concentrations in the control of *Alternaria* leaf blight in potatoes was determined in two consecutive years (2004, 2005). In both years first symptoms were observed on lower leaves three weeks after crop emergence. Although first early blight lesions were detected early in the growing season each year, disease severity remained at a low level (1-5%) until end of July, even in control (coversprayed) plots.

In the year 2004 in mid of August half of the canopy in the coversprayed plots was destroyed by early blight. Disease severity in the control plots then increased dramatically until August 25th when over 95% of the foliage were affected.

In 2005 the start of the early blight epidemic was later with the consequence that the AUDPC value in the control was less than the AUDPC value reported for 2004. In the beginning of September the disease level was 20%. Two weeks later over 95% of the foliage were infected by *Alternaria*.

The overall effect of fungicide concentration on early blight severity, as expressed by AUDPC, was highly significant (Table 1).

Table 1. Early blight control at different fungicide concentrations with mancozeb tested at Kirchheim 2004 and 2005. Fungicide efficacy is expressed as area under the disease progress curve (AUDPC).

	2004			**2005**		
	AUDPC[a]	**SC**[b]	**Tukey-B**[c]	**AUDPC**[a]	**SC**[b]	**Tukey-B**[c]
Control	2795.0	199.2	a	1876.7	69.5	a
MZ-500	2460.2	153.2	a,b	1754.2	65.2	a
MZ-750	2081.8	280.7	b,c	1722.2	55.9	a,b
MZ-1000	1818.7	195.3	c,d	1531.2	119.2	b,c
MZ-1500	1488.7	218.4	d,e	1421.8	35.9	c,d
MZ-2000	1473.7	189.5	d,e	1248.0	141.9	d
MZ-4000	1240.3	205.5	e	949.3	84.1	e

[a]Mean of area under disease progress curve.
[b]Standard deviation from the mean of 3 replications.
[c]Numbers of each column followed by the same letter are not significantly different at p= 0.05 according to the Tukey-B test.

All applicated fungicide concentrations were able to reduce disease severity as compared to the untreated control plot. However, mancozeb at lower concentrations was less effective than at higher concentrations. Significant differences in disease severity among different fungicide concentrations were observed until the end of the season. Treatments against early blight showed that disease severity was the lowest at fungicide doses of 4000 g mancozeb. In the year 2004 there was no further markable influence on disease progress at mancozeb concentrations above 1500 g per hectare. In that year the most effective treatments were at the 1500 g doses of mancozeb. On August 25th mancozeb at the reference dose rate and higher significantly suppressed early blight infection levels below those of the untreated control.

In 2005 the mean AUDPC for the reference mancozeb dose rate (1500 g mancozeb) was 1421. As in the year 2004 there was no significant difference among dose rate of 1500 g and 2000 g mancozeb. In both years there was no significant difference between the control and the plots treated with 500 g mancozeb. A reduction of 50% of the reference mancozeb dose rate (1500 g) results in a significant higher AUDPC.

Influence of plant age on disease severity of early blight

Disease progress curves pointed out, that potatoes which emerged earlier had the greatest severity in terms of leaf symptoms due to early blight compared to potatoes which emerged later. In average the late planted potatoes were about 12 to 14 days delayed in their growth compared to the earlier planted potatoes. The overall area under the disease progress curve was significantly greater for potatoes which had been planted earlier than for the late planted. Under field conditions early blight appeared at almost the same time in mid of June, regardless the crop age. The initial stage of disease development of early blight was similar in both years (2004, 2005) and stayed on a low level until end of July. Early blight disease did not increase until begin of August. Progress of disease development was similar for both planting dates, however with a temporal delay of around 10 to 14 days in the later planted potatoes (Figure 4). Whilst in 2004 the late-planted potatoes reached the disease severity of the early-planted potatoes, their disease development remained on a lower level in 2005 (Figure 5).

Discussion and conclusion

Over both sequent years in field trials, early blight has been a destructive disease and caused yield losses due to premature defoliation. Mancozeb was effective as protectant fungicide for the control of early blight when applied as a preventative spray. Reduced efficacy was observed when mancozeb was used with less than the reference dose rate. Fungicide treatments with less then 1000 g mancozeb were not satisfying in disease control. This loss of disease control of early blight indicates that fungicide treatments with mancozeb with less than 1000 g a.i./ha should not be encouraged. However, when environmental conditions were favourable for the development of early blight (2004), lower application rates of mancozeb (less than 1500 g) provided unsatisfactory control of the disease. The results indicate that the dose rates of mancozeb should be at least 1000g to reduce significantly early blight infections compared with the control plot.

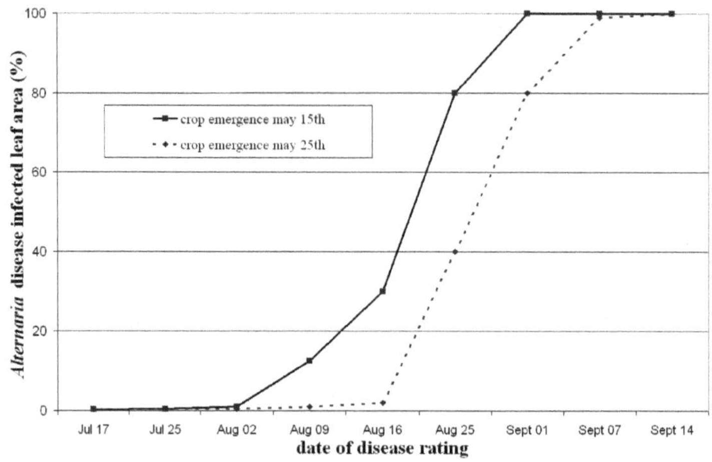

Figure 4. Disease progress of early blight 2004 in relation to crop emergence.

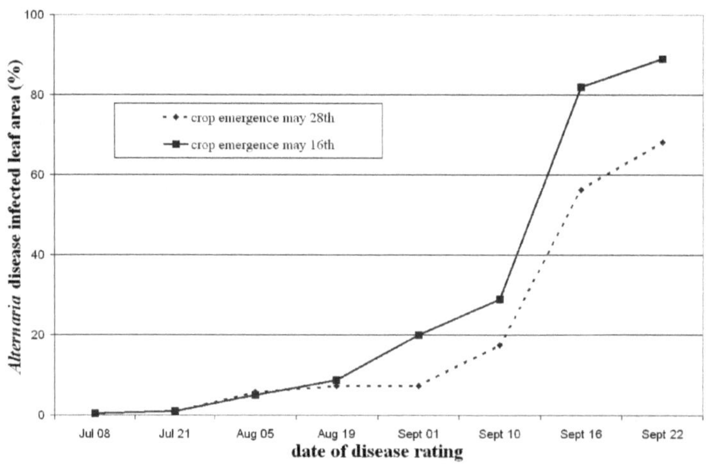

Figure 5. Disease progress of early blight 2005 in relation to crop emergence.

At the same time the plant age had a certain influence on disease incidence of early blight. These results show that at a fixed rating date, potatoes, which are in a later stage of maturity, are predominantly more affected by early blight as physiologically younger potato plants. However, early blight is known to be a disease that primarily affects physiologically old or weakened plants (Radtke *et al.*, 2000, Douglas and Pavek 1974). Despite the fact that young potato leaves can be infected by early blight the disease epidemic only bursts if a certain physiological age is exceeded. The fungus attacks most severely the oldest leaves while the youngest appear symptom less. Despite the lack of visible symptoms, many of these symptom less leaves already contain the pathogen (Rowell 1953) what can be proved by PCR technique

(Bahnweg *et al.*, 1998). Johanson and Thurston (1990) described this apparent immunity as a physiological mechanism that inhibits subsequent development of the pathogen in the plant. The physiologically younger potatoes remained healthy and photo synthetically active for a longer time which resulted in a higher yield.

References

Bahnweg, G., S. Schulze, E.M. Möller, H. Rosenbrock, C. Langebartels and H.Jr. Sandermann (1998). DNA Isolation from Recalcitrant Materials Such as Tree Roots, Bark, and Forest Soil for the Detection of fungal Pathogens by Polymerase Chain Reaction. Analytical Biochemistry 262: 79-82

Chelkowski, J. and A. Viskonti (1992). Alternaria-Biology, plant diseases and metabolites. Elsevier Verlag, Amsterdam

Douglas, D.R. and J.J. Pavek (1972). Screening potatoes for resistance to early blight. American Potato Journal 49:1-6

Ellis, M.B. and I.A.S. Gibson (1975). Alternaria solani. No. 475, CMI Descriptions of Pathogenic Fungi and Bacteria. Common Mycol. Inst. Kew, Surrey, England 2 pp.

Hausladen, H., E. Bässler and N. Asensio (2004). Early blight of potato. In: C.E. Westerdijk and H.T.A.M. Schepers (eds.), PPO Special Report no. 10: 173-177

Johanson, A. and H.D. Thurston (1990). The effect of cultivar Maturity on the resistance of potatoes to early blight caused by Alternaria solani. American Potato Journal Vol.67: 615-623

Johnson, K.B., E.B. Radcliffe and P.S. Teng (1986). Effects of interacting populations of Alternaria solani, Verticillium dahliae, and potato leafhopper (Empoasca fabae) on potato yield. Phytopathology Vol. 76, no. 10: 1046-1052

Petrunak, D.M. and B.J. Christ (1992). Gentis Isoenzyme Variability in Alternaria solani and Alternaria alternata. Phytopathology Vol 82: 1343-1347

Radtke, W., W. Rieckmann and F. Brendler (2000). Kartoffel: Krankheiten, Schädlinge, Unkräuter (potato: diseases, pests, weeds). publishing company Th. Mann, Gelsenkirchen-Buer

Rotem, J. (1994). The genus Alternaria; Biology and pathogenicity. American Phytopathological Society, St. Paul, Minesota

Rowell, J.B. (1953). Leaf blight of tomato and potato plants: factors affecting the degree of injury incited by Alternaria dauci f. solani. Agricultural Experiment Station Bulletin No. 320. Univ. of Rhode Island, Kingston

Venette, J.R. and M.D. Harrison (1973). Factors affecting infection of potato tubers by Alternaria solani in Colorado. American Potato Journal. 50: 283-292

Waals, van der, J.E., L. Korsten and T.A.S. Aveling (2001). A Review of early blight of potato. African Plant Protection 70 (2): 91-102

Waals, van der, J.E., L. Korsten and B. Slippers (2004). Genetic diversity among Alternaria solani isolates from potatoes in South Africa. Plant Dis. 88: 959-964

Curriculum vitae – Hans Hausladen

Hans Hausladen studied agriculture at the Technical University Munich (TUM) and finished with the master of agriculture in 1996. The master thesis was focused on the sensitivity of *Phytophthora infestans* populations to metalaxyl in Germany. After the study he worked as a PhD student for five years at the chair for phytopathology and received the PhD degree in 2002. During this time the main objective was the development and validation of a DSS (decision support system) for the control of late blight in potatoes. The model was introduced in practice in the year 2000. Since that time this DSS (PhytophthoraModell Weihenstephan) is the basis for the fungicide strategy in Germany. After he received the PhD degree he followed up this research projects in potato. At the moment he works in a post-doc position at the chair for phytopathology being responsible for teaching students and various research projects in cereal and potatoes. His research focuses on the fungi *Phytophthora infestans* and *Alternaria ssp.* and modelling of plant diseases.

Curriculum vitae – Jürgen Leiminger

Jürgen H. Leiminger was born in 1978. After he had acquired the general university-level graduation he studied agriculture, focused on plant science, at the Center of Life and Food Science Weihenstephan, which belongs to the Technical University Munich. His diploma thesis "Erhebungen zum Sporenflug von *Alternaria spp.* an Kartoffeln und Untersuchungen zur Wirkung neuerer Fungizide" was written at the chair for phytopathology Weihenstephan. This work focused on epidemic studies and effective disease control of early blight in potatoes. He then worked for KWS, which is one of the world´s leading plant breeding companies, as expert adviser in oilseed rape for Germany. After an offer of a PhD thesis he returned to the chair of phytopathology at the Life Science Center Weihenstephan. Since April 2005 he is now working on the topic of the potato disease early blight. There he especially focuses on the improvement of the disease control.

Beyond his studies Jürgen Leiminger attended several practical training at home and abroad. Apart from agricultural practical training on a farm he worked for the German state as well as for a crop protection company. During a three and a half months stay in South Africa he attended a training farm as supply teacher. Apart from this activity he was active in the execution of work within the range of development assistance. Beside his PhD work he still engages in projects for rural development projects in South Africa.

Company profile – Technical University of Munich: Life Science Center Weihenstephan

The Technical University of Munich (TUM) is one of the most well-known and famous universities of Germany. In more than 12 different faculties as well as numerous research stages promising research projects are bundled at the highest stage. The Life Science Center Weihenstephan (WZW) is one of the faculties of the TUM. The public research institutes of the Life Science Center are focused on the problems of healthy nutrition, creation and preservation of a natural environment. "Research for life and nature" is the motto of the institution. Interdisciplinary research and synergism through cooperation of the institutes are important features of the TUM, establishing the high international standard. The chair of phytopathology is part of the Life Science Center and one of eight chairs within the field of crop sciences. The central topics of phytopathology are to augment knowledge about host-pathogen interactions and diagnostics of plant diseases, as well as to estimate and control epidemiology of plant pathogens. The research, maintained by four working groups, is focused on various parameters from the molecular and biochemical level up to greenhouse and field tests. Plant diseases are a big problem resulting in yield loss and reduced quality. Important topics are disease management and modelling in order to prevent plant diseases and to save the crop at a high quality.

Effect of different forms and additional potassium fertilisation on yield and industrial quality of potatoes

K. Demeulemeester[1] and J. Bries[2]
[1]Interprovincial Research Institute for Potato Production, Ieperseweg 87, B-8800 Rumbeke-Beitem, Belgium
[2]Soil Service of Belgium, Willem de Croylaan 48, B-3001 Leuven-Heverlee, Belgium

Abstract

The results of six fertilising trials carried out in 2004 and 2005 on the cultivar Asterix in loam and sandy loam soils in Flanders are presented. The trials compare the effect of different forms of potassium fertilizers applied shortly before planting and during the growing season. The results demonstrate the possibilities of controlling black spot susceptibility and dry matter content before and during the growing season. The fertilisation resulted in a yield response of 4.0% and the proportion of large tubers rose from 72.6 to 77.0% when K was applied. The form of potassium had no significant impact. No negative impact could be observed from high amounts of chloride potassium. Dry matter content and black spot susceptibility decreased clearly when potassium chloride was applied, both before planting and during the growing season. K fertilizers had a positive influence on fry colour, but the impact of the form was rather limited, with a small advantage for potassium chloride. After 5 months of conservation these differences became even less pronounced. When no K was applied the uptake of the tubers was above 300 kg K_2O/ha. A basic K supply before planting resulted in an extra export of about 14 kg/ha whereas the export rose only by 5 kg K_2O/ha after the additional supply during the growing season.

Keywords: potassium, chloride, black spot, dry matter, fry colour

Introduction

In Belgium most of the potatoes are cultivated for industrial processing, especially for French fry production. In 2005 the use of potatoes for processing has risen to more than two million tons, according to figures published by the Belgian trade association Belgapom. In spite of some agronomic disadvantages - little resistance against several diseases, sensitivity to secondary growth and moderate dry matter content when cultivated on sandy loam soils - Bintje remains by far the most widely grown cultivar. Nevertheless some alternative, high yielding cultivars have proved themselves the past years, for example cvs. Asterix, Saturna, Felsina, Fontane and Ramos. According to a survey conducted in April 2006, Asterix represents 10% of the ware potato area in Flanders (Clarysse, 2006). For Saturna and Felsina the share was estimated with 6% and 4%, respectively, of the ware potatoes area.

Besides the potential of a high yield, some of these cultivars also have a high dry matter content, for instance Asterix. Although this characteristic is favourable to obtain a good processing quality it implies an increasing risk on black spot and consequently more risk on refusal for processing purposes.

This article describes the research on the influence of potassium fertilisation on yield, dry matter content, black spot susceptibility and fry colour for the cultivar Asterix. The results of the 2004 and 2005 trials in (sandy) loam soils are presented here.

Methodology

In total 6 trials were conducted on sandy loam (3), and loam (2) soils in the provinces East and West-Flanders. The cultivar grown was Asterix (Table 1).

Different form and additional potassium regimes were compared in randomized block design (split-plot) with 4 replicates: 1 regime without potassium fertilizers, 3 regimes with potassium chloride, 3 regimes with potassium sulphate and 3 regimes with a mixture of potassium chloride and sulphate (50/50). Minimum size plot was 9 x 9m (Figure 1).

Nitrogen fertilisation was performed shortly before planting according to the field and cultivar specific recommendations obtained from the N-index method of the Soil Service of Belgium (Bries *et al.*, 1995; Hofman and Salomez, 2000). Potassium fertilisation has been based on a standard soil analysis by the Soil Service of Belgium. The advised dose was applied at most 3 weeks before planting (both sulphate and chloride forms). In some regimes an additional supply of 120 kg K_2O, both chloride and sulphate potassium, was applied shortly before flowering (around mid June). Only mineral fertilizers were used, both for nitrogen and potassium.

Yield was measured at harvest on minimal 9 m² for sizes below 35 mm, 35-50 mm and above 50 mm. Underwater weight was assessed with a feculometer on 50-55 mm sized tubers. Fry colour was determined according the VBNA-method (Munsell Color Company colour chart) on 20 tubers per plot.

Table 1. Main characteristics of the trial fields.

Location	St.-M.-Lierde (2004)	Poperinge (2004)	Zonnebeke (2004)	St.-M.-Lierde (2005)	Poperinge (2005)	Zonnebeke (2005)
Soil type	loam	sandy loam	sandy loam	loam	sandy loam	loam
Previous crop	maize	red cabbage	sugar beet	peas	sugar beet	winter wheat
K-content soil (mg/100g)	16	27	28	22	32	27
Fertilisation (mineral - kg/ha)						
N [a]	194	147	152	129	96	100
K_2O [a]	280	200	180	230	190	220
basis (= advice)	6 Apr	13 Apr	9 Apr	13 Apr	17 May	1 Apr
extra dose (120 units)	4 Jun	11 Jun	10 Jun	10 Jun	24 Jun	12 Jul
Planting date	14 Apr	15 Apr	26 Apr	24 Apr	18 May	18 May
Harvest date	29 Sep	29 Sep	5 Oct	11 Oct	5 Oct	12 Oct

[a] advised dose

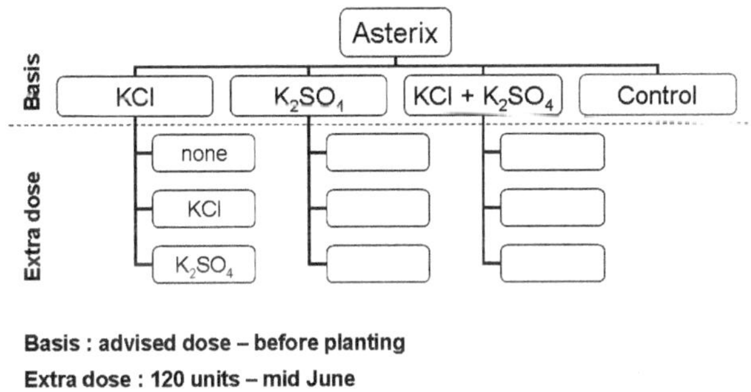

Figure 1. Set-up of the trials on potassium fertilisation in cv. Asterix (2004 and 2005).

Black spot susceptibility was measured on 50 manually harvested tubers (size 50-55 mm) per plot. After being stored for 2 days at 8°C, tubers were shaken during 1 minute and stored for 3 days at 16°C to enhance the enzymatic discolouration. At the final assessment tubers were peeled for 80% and arranged in 5 classes according to the size of the black spots (adapted ATO-method). Both potassium and chloride content of the tubers were measured from the dry matter: potassium is assessed via ICP after extraction with nitric acid whilst chloride content is determined in a continuous flow of an aqueous extract. Although Van Canneyt *et al.* (2004) reported fluctuations in K content of the tubers in pre-harvest and harvest period, the K and chloride content were only assessed at harvest.

Finally, tubers were stored at 8°C until March for evaluating the changes in fry colour during storage. Germination was inhibited by application of CIPC (DP, 2 kg per tonne). Only tubers from the trials in Lierde and Zonnebeke were stored (as well in 2004 as in 2005).

Results and discussion

When no potassium was supplied to the potato crop, a total yield of 62.5 t/ha was observed (Table 2). The marketable yield (+ 35 mm) was 60.2 ton/ha. When K was applied before planting, total yield increased by 2.5 t/ha which corresponds with a yield response of 4.0%. A similar response on soils with high K content (> 20 mg/100 g) is reported by Perrenoud (1993), *i.e.* 3.8% when at least 120 kg K_2O/ha were applied. Tuber size was also affected by the potassium: the marketable yield rose from 96.3 to 98.1% as K application was increased from nil to the advised dose. The proportion of large tubers (over 50 mm) rose from 72.6 to 77.0% when K was applied. When the form of potash fertilizer is considered, application of chloride seems to enhance larger tuber size but this difference is not significant. Similar observations are quoted by Perrenoud (1993).

When additional potash was applied in June, yield increased slightly when K_2SO_4 is used. The increase became even significant when KCl was applied (total yield).

Consequently these trials indicated clearly that an application of chloride shortly before planting had no negative effect on the yield or tuber size for cv. Asterix.

Table 2. Yield (average figures for 2004 and 2005).

Yield t/ha	+35 mm		35-50 mm		+50 mm		Total	
Control (no K_2O)	60.2		14.8		45.4		62.5	
Basic K-supply (before planting)								
KCl	62.7	a	12.9	a	49.8	a	63.9	a
K_2SO_4	62.8	a	13.9	a	48.8	a	64.1	a
KCl & K_2SO_4	63.2	a	13.6	a	49.6	a	64.5	a
Extra K-supply (June)								
none	62.6	b	13.6	a	48.9	a	63.9	b
KCl	63.3	a	13.1	a	50.2	a	64.5	a
K_2SO_4	62.7	ab	13.7	a	49.1	a	64.0	ab

Values in the same column followed by the same letter are not significantly different (P 0.05)

However, in practice, the chloride form is applied during winter to avoid negative effects of chloride on yield. This apprehension is therefore unfounded in the (sandy) loam soils of Belgium. Similar conclusions are made by Veerman (2001) for the Netherlands (clay soil). Previous research however indicated that in some years potassium chloride application shortly before planting can reduce yield. Bries and Vanden Auweele (1999) found that potassium chloride resulted in a yield decrease of about 10% in 1997 and 1998 compared with potassium sulphate whereas in 1999 yield was 3.5% higher with chloride. These trials were performed in sandy loam soils on Bintje with a potassium supply of 250 - 320 kg K_2O/ha.

Although the great difference in total yield and calibration between the two years, no interaction could be observed between year, basic K-supply or extra K-supply. Due to the late planting in 2005, total marketable yield only reached 55.3 t/ha, which is very low compared to the 70.4 t/ha in 2004. Also the proportion of large tubers (over 50 mm) was noticeably inferior in 2005 (70.8% in 2005 *vs.* 84.7% in 2004).

In spite of different seasons and growing conditions, similar values were observed for the underwater weight in both years. Application of potash sulphate before planting resulted in a decrease of 14 g/5 kg underwater weight (Table 3).

When sulphate was replaced by chloride, the underwater weight decreased even more than 20 g. Half a dose of chloride and half a dose of sulphate also affected clearly the underwater weight. The effect of potassium and form of potassium on dry matter content as observed in our experiments is also reported by Perrenoud (1993) and Veerman (2001). An additional dressing of potash in June had minimal impact when sulphate was used. Contrarily, applying chloride clearly resulted in an obvious decline of dry matter content.

When black spot susceptibility was considered, very high values were observed when the crop's potassium was denied. Application of K before planting resulted in clearly lower values, especially when the chloride form was used. Nevertheless, the differences between the different

Table 3. Industrial quality parameters at harvest (average figures for 2004 and 2005).

Parameter	Underwater weight		Black spot susceptibility		Fry colour		Heterogeneous fries	
Scale	(g/5 kg)		(0-400)		(0-6)		(%)	
Control (no K_2O)	420		180		3.13		34	
Basic K-supply (before planting)								
KCl	396	b	129	a	2.94	b	24	a
K_2SO_4	406	a	149	a	3.02	a	27	a
KCl & K_2SO_4	398	b	133	a	2.98	ab	25	a
Extra K-supply (June)								
None	406	a	154	a	3.02	a	27	a
KCl	392	b	115	c	2.94	a	23	a
K_2SO_4	402	ab	141	b	2.98	a	26	a

Values in the same column followed by the same letter are not significantly different (P 0.05).

forms of potash were not significant. An extra dressing in June reduced black spot susceptibility significantly, not only with sulphate but especially after chloride application.

Fry colour improved when the recommended dose of K was applied before planting. Although the improvement was limited, replacing the sulphate form by chloride gave better results. The mixture of chloride and sulphate scored in between. An additional dose of 120 kg K_2O in June seemed to improve fry colour, with the largest gain after chloride application, but the obtained differences were not significant (P 0.05). As on percentage heterogeneous fries, K-fertilization decreased the number of heterogeneous fries with a slight advantage for the chloride form, but none of the differences proved to be significant.

According to Vertregt (1968, quoted by Perrenoud, 1993) tubers with a K content below 2% are in general very susceptible to black spot while the incidence is nil or very slight at tubers content above 2.5% K. In our trials tuber content was about 1.8% when no K was fertilized, which resulted in highly susceptible tubers. Applying K shortly before planting improved tuber K content, however very limited (+ 0.2%). No effect of the form could be seen. An extra supply in June resulted in a higher K content when the chloride form was used while sulphate had only a minimal impact (Table 4).

Chloride content was strongly correlated with the form of potash. Even when only K_2SO_4 was applied, chloride content of the tubers was higher than the plots which did not received any K-fertilizer. Tuber content did not differ between years for K but for chloride the difference was significant (see Table 5).

Multiplying yield with K content of the tubers gives an idea of the mineral uptake or export from the soil. Since the foliage stays on the field, this factor is not taken into account. In spite of an average K-supply of 217 kg K_2O/ha, a basic dressing of potassium before planting

Table 4. Mineral tuber content and export from the soil (average figures for 2004 and 2005).

	Tubers - mineral content (mg/100 g DM)				Soil - mineral export (kg/ha)			
	Potassium		Chloride		Potassium		Chloride	
Control (no K_2O)	1,802		210		315		28	
Basic K-supply (before planting)								
KCl	2,007	a	360	a	328	a	49	a
K_2SO_4	1,953	a	244	c	328	a	34	b
KCl & K_2SO_4	2,008	a	305	b	331	a	42	ab
Extra K-supply (June)								
None	1,949	b	279	b	326	a	39	b
KCl	2,031	a	363	a	331	a	49	a
K_2SO_4	1,987	ab	267	b	331	a	37	b

Values in the same column followed by the same letter are not significantly different (P 0.05)

led to an export of 329 kg K_2O/ha, which was only 14 kg higher than the non K fertilized plots. An additional supply of 120 kg K_2O/ha in June resulted in an increase of the export with only 5 kg/ha. No effect of the form could be seen. The chloride export from the soil has been strongly correlated with the potassium form and therefore was higher for chloride. As a result of the great variation in yield between years, potassium and chloride export from the soil was significantly smaller in 2005 (Table 5).

Finally the evolution of the fry colour during conservation was evaluated (Table 6). For this purpose only samples from the locations of Lierde and Zonnebeke were stored, both in 2004 and 2005. At harvest, no statistical differences could be observed between the K fertilized plots although the tendency for better fry colour after potassium (chloride) application seemed clear. Also at the assessments of January and March, best fry colours were obtained after application of potassium chloride (both basic and additional dressing). However when the extra K-supply in June is considered, the beneficial effect of the chloride form on the fry colour seems to be reduced after 5 months storage. When no additional K was applied, fry colour deteriorated more slowly.

Conclusions

On the basis of the trials conducted in 2004 and 2005 on Asterix, one can conclude that the form of potassium plays a major role on the industrial quality, especially on dry matter content and black spot susceptibility. An additional dressing with potassium chloride in June had a clear impact on both parameters. The influence on fry colour was rather limited. Potassium fertilization resulted only in a small yield increase, which is undoubtedly correlated with the moderate to high K content of the soil. No negative impact on yield was observed from the chloride form applied shortly before planting or during the growing season. Although black

Table 5. Influence of the year on various parameters (average K-fertilized plots).

Year		2004		2005	
Trial characteristics					
N - advised dose	kg/ha	164		108	
K_2O - advised dose	kg/ha	220		213	
Growing days		166		152	
Yield					
+ 35 mm	t/ha	70.4	a	55.3	b
35-50mm	t/ha	10.8	b	16.1	a
+50 mm	t/ha	59.6	a	39.2	b
Total	t/ha	71.4	a	56.9	b
Industrial quality parameters					
Underwater weight	g/5 kg	399	a	401	a
Black spot susceptibility	0-400	129	b	144	a
Fry colour	0-6	2.95	b	3.01	a
Heterogeneous fries	%	25	a	26	a
Mineral tuber content					
Potassium	mg/100 g DM	1,990	a	1,988	a
Chloride	mg/100 g DM	317	a	289	b
Mineral export from the soil					
Potassium	kg/ha	370	a	288	b
Chloride	kg/ha	49	a	34	b
Fry colour - evolution					
Δ Oct - Jan		0.36	a	-0.07	b
Δ Oct - Mar		0.15	a	0.05	b

Values in the same row followed by the same letter are not significantly different (P 0.05)

Table 6. Evolution of the fry colour (2 locations per year; average figures for 2004 and 2005).

Fry colour (0-6)	October	January	March	Δ -Oct - Jan	Δ -Oct - Mar
Control (no K_2O)	3.11	3.25	3.16	0.14	0.05
Basic K-supply (before planting)					
KCl	2.87 a	3.03 b	3.02 a	0.16 a	0.15 a
K_2SO_4	2.97 a	3.10 a	3.05 a	0.13 a	0.08 a
KCl & K_2SO_4	2.95 a	3.08 a	3.02 a	0.14 a	0.07 a
Extra K-supply (June)					
None	2.99 a	3.09 a	3.06 a	0.10	0.07 b
KCl	2.88 a	3.01 a	3.00 a	0.14 b	0.12 a
K_2SO_4	2.92 a	3.11 a	3.03 a	0.19 a	0.11 a

Values in the same column followed by the same letter are not significantly different (P 0.05).

spot susceptibility decreased clearly, only a limited effect on the K content of the tubers was observed. This implicates that only a minor part of the applied K fertilizers is exported from the field at harvest. Applying additional potassium in June appears to reduce the risk on high black spot appearance, but should be reserved for those cultivars with high dry matter content and high black spot susceptibility. However, trials are continuing to find out more about the effect of form and timing of potassium fertilizer on Asterix grown in clay soils and on other cultivars with high dry matter content.

Acknowledgements

This project was funded by the Flemish Ministry of Agriculture and implemented in the framework of "Landbouwcentrum Aardappelen" (agricultural centre for potatoes). The tuber analyses were supported by K+S Benelux. Special thanks also to our partners for the work on the trial fields, *i.e.* the Section Vulgarisation of the Flemish Ministry of Agriculture and the Agricultural Institutes of Poperinge and Sint-Niklaas.

References

Bries, J., Vandendriessche, H. and Geypens, M. (1995). Bemesting en beregening van aardappelen in functie van opbrengst en kwaliteit. (Fertilisation and irrigation of potatoes in function of yield and quality). IWONL, Brussels, Belgium, 250 pp.

Bries, J., Vanden Auweele, W. (1999). Keuze van de kalimeststof. (Selection of the potassium fertilizer). In Aardappelen - Onderzoek en voorlichting 1999 in België. (Potatoes - research and vulgarisation 1999 in Belgium), 67-68.

Clarysse, L. (2006). Areaal bewaaraardappelen stijgt 5%. "Area ware potatoes rises 5%". PCA Newsletter, 14 (6), 4-6.

Hofman, G. and Salomez, J. (2000). Nitrogen decision support systems in potato production. D.K.L. MacKerron and A.J. Haverkort (Eds.). Management of nitrogen and water in potato production. Wageningen Academic Publishers, 219-232.

Perrenoud, S. (1993). Fertilizing for high yield potato. IPI Bulletin 8 (2nd revised edition). Basel, Switzerland, 94 pp.

Van Canneyt, T., De Temmerman, L. & Sonck, B. (2004). Fluctuations in potassium content of mature potato tubers in the pre-harvest and harvest period. Meeting of EAPR, Prague, Czech Republic, April 19-23, 71-77.

Veerman, A. (2001). Variation in tuber quality between and within lots of ware potato cultivars. In Dutch, PhD thesis, Wageningen University, The Netherlands, 253 pp.

Vertregt, N. (1968). Relation between black spot and composition of the potato tuber. European Potato Journal, 11, 34-44.

Curriculum vitae – Kürt Demeulemeester

Kürt Demeulemeester studied agricultural sciences at the Catholic University of Leuven (KUL) in Belgium, from which he received the Agronomic Engineer degree in 1996. In 1997 he joined the Interprovincial Research Institute for Potato Production (PCA vzw), department Kruishoutem (province of East Flanders), as crop adviser and researcher. He worked on projects for integrated potato culture and coordinated the applied research and extension on potatoes in Flanders. The main themes of his research were fertilisation, conservation and varieties.

In 2003 he transferred to the department of Beitem (province of West Flanders) where he became responsible for this division. In addition to the mentioned themes he also started working on seed potato production. Since 2005, he is secretary of the Flemish association of seed potato growers.

Company profile – Interprovincial Research Institute for Potato Production (PCA vzw)

The Interprovincial Research Institute for Potato Production (PCA) is an independent, non-profit-making information and research centre for potato growers and the potato sector. The PCA was established in 1992 as a joint initiative of the provinces of East and West Flanders, which explains why there are two offices, in Beitem and Kruishoutem (Belgium). At the moment the PCA has 1000 members spread throughout Flanders, representing approximately one-third of the total cultivated acreage.

The quality analyses are carried out at the laboratory of Beitem while extension is concentrated in Kruishoutem. Both departments have their own trial fields. Trials performed in accordance with local conditions are focused on aspects such as plant production (base material, leaf control), varieties for early and ware potatoes, potatoes for the fresh market and the industry, fertilisation, cultivation techniques (ridging, planting distance, irrigation), plant protection (diseases, insects, weeds), storage. The PCA provides individual counselling with regard to cultivation techniques, plant protection, storage and crop recording.

Thanks to the wide range of services it offers, the PCA contributes to sustainable and profitable production of quality potatoes. The cornerstone of its services is the late blight warning system. Using 32 automatic observation posts, potato blight can be properly controlled. Information is communicated to members mainly by fax or mail. Recently a new website has been developed (www.pcainfo.be).

Calcium in potato growing

R.A. Bosch and R.A.J. Velema
HLB, Kampsweg 27, 9418 PD Wijster, the Netherlands

Abstract

The occurrence and severity of diseases in plants is often affected by the nutrient availability. Also in potato growth there are some interesting interactions. In this paper the interactions related to calcium are reviewed.

Keywords: calcium concentration, soil, growth

Introduction

Calcium deficiency causes several symptoms. In seed tuber, sub-apical necrosis of sprouts is a typical symptom. Due to the deficiency of calcium, sprouts die off and when calcium is applied to the tuber or becomes available to the plant new sprouts appear. It is believed that this phenomenon can repeat itself one or more times. Symptoms of calcium deficiency in seed tuber are often confounded with the symptoms caused by *Rhizoctonia solani*. Results of experiments with tubers with different calcium concentration show that tubers with high calcium give in general more stems per plant (Velema *et al.*, 2001a).

In consumer potatoes calcium deficiency can cause brown spots in tubers. These tubers are not marketable anymore. Brown spots from calcium deficiency are sometimes confounded with symptoms caused by *Trichodoridae*.

Calcium deficiency may induce tuber rot cause by *Erwinia* (Kelman *et al.*, 1993). Zwart *et al.* (1999) reported a higher infestation of *Fusarium solani* at lower tuber calcium concentrations. *F. sulphureum* was not affected. This relationship is presented in Figure 1.

Calcium fulfils and important role in potato growth. It is an element which makes cell structure and membranes stronger. Especially, the role in the pectin complex. Some disease cause enzymatic destruction of pectin by means of the enzyme pectinase.

Soil calcium concentration

First research on calcium at HLB Wijster (Netherlands) took place in 1995 (Mulder *et al.*, 1995). The reason arose from the observed difference in quality of seed tuber from clay soil soils compared to those from sandy soils. The sprouts of seed tubers from clay grew better. An inventory of 27 samples indicated that the difference in seed tuber quality was strongly related to calcium concentration of the tuber. Tubers from clay had a mean calcium concentration of 15 mg Ca per 100 g fresh tuber, whereas the mean calcium concentration of tubers from sandy soil was about 5 mg Ca per 100 g fresh tuber. Other elements (N, P, K, Mg, S, B, Mn and Zn) were not affected by soil type.

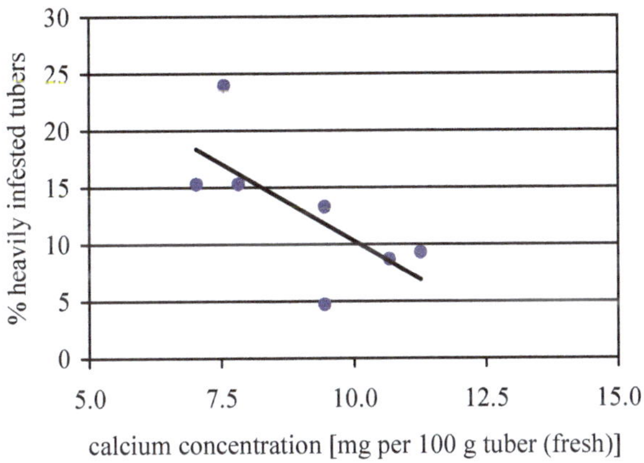

Figure 1. The relationship between tuber calcium concentration (cv. Karnico) and infestation due to Fusarium solani *(% heavily infested tubers = 37.3 - 2.70 * calcium concentration; R^2=0.49), Zwart* et al., *1999.*

Compared to those from sandy soils the calcium concentration of clay soils was far much higher. In general calcium concentration of sandy soils was higher as organic matter concentrations increased. This is also established by Zwart *et al.* (1999).

On sandy soils calcium concentrations are higher at increasing pH levels (Mulder *et al.*, 1995). This is probably due to the effect of lime. It increases both pH and calcium concentrations.

Cultivar

There are also differences in tuber calcium concentration between cultivars (Velema, 1998). The cultivar with the lowest tuber calcium concentration was Elles (± 4.5 mg Ca per 100 g fresh tuber), Kardent had the highest concentration (± 8 mg Ca per 100 g fresh tuber). It was often demonstrated that practically all potato growers know that Kardent showed less problems during storage.

Fertilisation

It seems not easy to increase tuber calcium concentration. Probably because calcium is not easily (re)transported within the plant. Kratzke and Palta (1986) suggested that calcium uptake not only took place by the root, but a main part of the calcium is taken up through the peel of the newly formed tubers. Experiments with several dosages and fertilisers incorporating techniques leads to the conclusion that tuber calcium concentration only increased when calcium concentration in the root zone increased (Velema, *et al.*, 2001b).

There is a log linear relationship between concentration of applied gypsum to the soil and the calcium concentration in the tuber (Figure 2). Experiments with boron to increase calcium concentration in tubers was not found (Velema en Van de Griend, 2001).

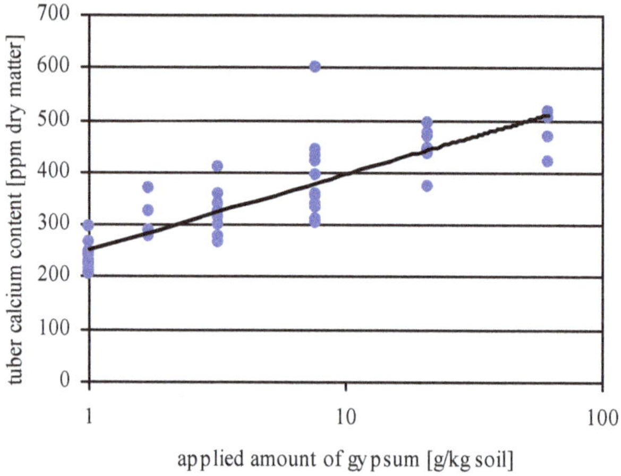

Figure 2. The relationship between the applied amount of gypsum and tuber calcium concentration. Tuber calcium concentration=252 + 62.9 x ^{e}log(applied amount of gypsum); R^2=0.71.

Discussion

In general, it can be stated that calcium deficiency plays an important role in the starch potato, consumer potato and seed potato production. Furthermore, there is a relationship between tuber calcium concentration and the occurrence of certain diseases. This relationship is strongly influenced by other production factors, *i.e.* water stress, heat, *etc.* Kelman *et al.* (1993) states that increasing calcium concentration of the soil lead to an increment of tuber calcium concentration. He also discussed the application rate and method, *i.e.* calcium should be readily available at maximal tuber bulking time. Experiments carried out by HLB do not show these trends, to increase the tuber calcium concentration from 250 to 500 ppm, almost 100 grams of gypsum per kilogram soil was needed. The difference might be an effect of different environmental conditions between the American and Dutch trial sites, for example soil moisture, humidity and temperature. Figure 3 shows the relationship between the amounts of calcium taken up during the growing season and tuber calcium concentration in time. It shows that less then three kg calcium per hectare was absorbed by the potato plant. This shows that only a relatively small amount of calcium is taken up by the tubers. This corroborates the findings of Kelman (1993).

Conclusion

The calcium concentration of the tuber has an influence on the occurrence of diseases. The higher the calcium concentration, the lower the disease severity. In general, seed potatoes cultivated on a sandy soil are more susceptible to bacterial soft rot then potatoes grown on a clay soil. In practice it is not easy to increase the tuber calcium concentration. However, symptoms and diseases caused by the calcium deficiency may, to a certain extent, be minimised by use of calcium fertiliser.

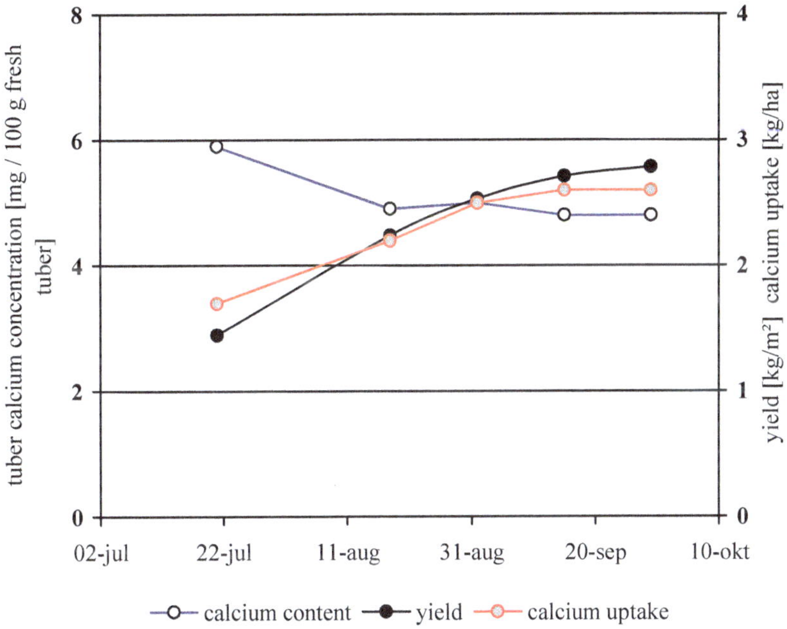

Figure 3. The relationship between the calcium uptake during the growing season and tuber calcium concentration.

References

Kelman, A., R.G. Mcguire and K-C. Tzeng, 1993. Reducing the severity of bacterial soft rot by increasing the concentration of calcium in potato tubers. Soilborne Plant Pathogens: Management of diseases with macro- and microelements. St Paul, Minnesota USA. APS Press (1993), p. 205-217

Kratzke, M.G. and J.P. Palta, 1986. Calcium accumulation in potato tubers: role of basal roots. Hort. Science 21 (1986), p. 1022-1024

Mulder, A., R.A.J. Velema and G. Veninga, 1995. Pilot experiments with foliar nutrition in the potato crop. HLB Wijster, Rapport 9520, 12 p.

Turkensteen, L.J. en A. Mulder, 1995. Beschikbaarheid voor het aardappelgewas van macro- en micro-elementen in relatie tot opbrengst en gevoeligheid voor ziekten. SIO Onderzoek 1995 (1995), p. 235-245

Velema, R.A.J. en G. Veninga, 1996. Oriënterend onderzoek in aardappelen naar de relatie tussen het gehalte aan hoofd- en spoorelementen in de knol en de knolkwaliteit. HLB Wijster. Rapport 9625, 5 p.

Velema, R.A.J., 1998. Calciumvoorziening in pootgoed. PPO Onderzoek 1998 (1998), p. 94-96

Velema, R.A.J. en P. van de Griend, 2001. Het effect van toediening van calcium en borium aan de grond op de opbrengst, kwaliteit en minerale samenstelling van pootaardappelen, Agrobiokon, PRI nota 101, 10 p.

Velema, R.A.J. en P. van de Griend, 2001. Het effect van het calciumgehalte van pootgoed op de ziektegevoeligheid (II), HLB rapport 118, 17 p.

Velema, R.A.J., P. van de Griend en H. Velvis, 2001a. Het effect van calciumgehalte van pootgoed op de groei en opbrengst van zetmeelaardappelen, Agrobiokon, PRI nota 104, 14 p.

Velema, R.A.J., P. van de Griend en H. Velvis, 2001b. Onderzoek naar de verhoging van het calciumgehalte in aardappelknollen, Agrobiokon, PRI nota 102, 8 p.

Zwart, K.B., H. Velvis, R.A.J. Velema and P. van de Griend, 1999. Micronutrienten en biotische en abiotische stress in zetmeelaardappelen. PPO Onderzoek 1999 (1999), p. 134-142

Curriculum vitae – Ronald A. Bosch

Ronald A. Bosch has studied plant production in the tropics at Larenstein International agricultural college and continued his studies at Wageningen University in the area of phytopathology (epidemiology/ecology). Afterwards he worked for almost ten years in the tropics, *i.e.* Latin America (Costa Rica), Caribbean (Windward Islands) and West Africa (Ghana). Beside this he worked at Larenstein International College in Deventer (The Netherlands). In general, his work entailed crop protection, agronomy, plantation management and biological disease control. He has frequently travelled around the world as a consultant for Fair Trade fruit producers. The last three years he works for HLB (The Netherlands) as a scientific researcher. At HLB he is responsible for the diagnostic services, molecular research, compost research, biological control, organic pesticides and precision agriculture.

His present research work is focused on potato wart disease (*Synchytrium endobioticum*), bacterial rot (*Pectobacterium* spp.) and is involved with research on precision agriculture, namely the relationship between pH and yield in a particular field. Besides this he becomes more involved in research on biological activity in the soil, how the manage it, the use of compost and the part it can play to manage nematode populations in the soil (biofumigation). The same is more or less valid to control soil pathogens by means of biofumigation of the soil in potato and sugar beet cultivation.

Company profile HLB

How to get the scientific information to the farmer in such a way that he/she can implement it and apply it within the present setting of the farm. That is the main objective form HLB, company that is officially started as HLB in 1999 and was before this time known as Hilbrands laboratory for soil diseases. In earlier days HLB its work area was emphasized on potato cysts nematodes in starch potatoes. At the moment, there are 19 people working with HLB, and its area of interest has increased to research in free living nematodes and cysts nematodes in potatoes, sugar beets, grasses, flower bulbs and tree nurseries. Since two years HLB started its molecular research program and that has already resulted in three commercial products. HLB is involved in the breeder research programs, namely the testing of resistance for quarantine organisms: potato cyst nematode and potato wart disease and is also the producer of inoculum for the two diseases. HLB is testing many nematicides and fungicides for chemical companies in potatoes, sugar beets, carrots, cereals and flower bulbs. Next to Phytophthora research, other fungi are also tested for example Alternaria, Fusarium and Verticillium. Another area of research is storage problems with potatoes (*Fusarium, Phoma, Rhizoctonia, Erwinia*). HLB is conducting trials with nutrients supply, foliar fertilisation and the relationship between nutrients and pests and diseases. The trial fields from HLB are located throughout the Netherlands at several farms, HLB itself has two laboratories (quarantine facilities with modern diagnostic services for farmers and intermediates, *i.e.* PCR en ELISA) and glass houses.

HLB conducts resistance trials for potato breeders. In these trials, the new varieties from the breeders are tested for resistance against Globodera spp. and *Synchytrium endobioticum*. Other disease can also included in these tests, *i.e. Fusarium* spp., *Helminthosporium* spp., *etc.* The results of most of the trials (except the confidential trials) is presented in the many different study groups from HLB, courses (within Europe) and via the website (www.hlbbv.nl) or magazines.

Product quality and certification

Potato quality: Getting the basics right

A. Veerman and R. Wustman
Wageningen University & Research Centre, Applied Plant Research (PPO), P.O. Box 430, 8300 AK, Lelystad, The Netherlands

Abstract

Because of high input, potato growing is expensive. Therefore it is very important, to minimize the risk of failure yield and quality. This is only possible by having available a basic level of inputs, knowledge and (local) information. From this level a permanent effort is necessary searching for the weakest link, improving it and moving on to the next weakest link. This paper describes the most important basics for potato production that have to be covered in order to be sufficiently assured of acceptable yield and quality. The chain of input and management starts with the seed that has to be of an appropriate cultivar for the market that is aimed for and it has to be both pathologically and physiologically healthy. The soil has to be of sufficient quality by texture, by absence or low levels of diseases and pests and by sufficient nutrient levels and it has to be tilled the right way and especially at the right time. Water and nutrient supply are crucial to both yield and quality of the crop. Only few areas in the world have consistent and sufficient natural precipitation to ensure unrestricted growth. The potato crop is sensitive to attack by a number of pests and diseases. In almost all potato growing areas potato crops are attacked by early or late blight or even both. It takes a considerable input of fungicides, equipment and knowledge to both maximize the control of disease and to minimize cost and environmental burden. Having completed the growing season is only half of the journey: tubers have to be harvested and stored with sufficient care because potatoes are susceptible to different types of damage which may result in loss of quality or spoilage due to subsequent disease attack. Many potatoes spend more time in storage than in the field. Desired storage conditions are easier to establish than field conditions. However, to do so a lot of hardware and software and skills to operate stores are needed to complete a successful delivery of potatoes that meet their demands.

Keywords: seed, soil, irrigation, crop protection, harvest, storage

Introduction

Growing potatoes is an expensive business: a lot of expensive input is required. It is also an intensive business: it not only requires material input, but also a permanent input of attention, information and knowledge. The latter is essential to make all the material input into a success and to produce a high quality crop which brings return on investment.

This means that all the input has to meet certain standards to sufficiently minimize the risk of failure with respect to both yield and quality level. From there it is a matter of continuously looking for the most limiting factor, improving it until another factor becomes limiting. Being able to apply this procedure implies that knowledge about all the basics is present, more or less chronologically covering: seed quality, fertilization, soil preparation, planting and ridging, irrigation, crop protection, haulm killing, harvesting, store loading and storage.

To a large extent quality characteristics have to be leading for the cultivation and storage management. In a free market producers have to compete with other suppliers. Fulfilling the quality demands of the buyer is crucial to financial return and the continuity in business relations. Maximizing the production within the limits of quality demands is essential for financial return as well, but is primarily the concern of the producer and not of the buyer.

Choice of cultivar

Year to year varying growing conditions have large consequences for most quality characteristics like dry matter content, reducing sugar content, tuber malformation, internal defects *etc.* The quality characteristics falling within specifications are essential to the marketability of the crop and require a lot of knowledge and skills regarding cultivation of potatoes. A very important measure is the choice of cultivar. The characteristics of cultivars are relatively stable and predictable and provide a solid basis for sufficient certainty of fulfilling specifications. In a highly optimized cultivation of potatoes within factors that are under the influence of the producer, the choice of cultivar is responsible for a large part of predictability of both quality characteristics and yield (Veerman, 2001; Veerman, Struik & Van Loon, 2002).

Cultivars interact with soil type and fertilization and other factors. Also some characteristics have to fit to local conditions like temperature, day length and length of the growing season. Given these facts it generally takes a few years of testing a newly introduced cultivar to find out its actual performance under local conditions (Veerman, 2001).

Seed quality and seed preparation

High quality seed can only be the result of specialization on its own. Growing high quality seed asks for adjusted planting rates, crop protection, haulm killing, harvesting and storage.

Health in terms of absence of diseases is one of the priorities for seed to be used for ware potato crops. Historically, seed production strategies and multiplication schemes were designed to control virus diseases which already proved to be effective when viruses were not even yet known as the cause of yield reduction. Depending on the virus and the proportion of secondary diseased plants yield may be reduced up to 50%. With help of spraying mineral oil and insecticides and with the help of monitoring of virus transmitting aphids to advise on timely haulm killing, virus diseases can be controlled quite well in areas with limited development of aphid populations. In general these are areas with a relatively cool climate.

A second major category of diseases are bacterial diseases which in general are easily spread in the clonal multiplication of potato. Some bacterial diseases have a quarantine status in parts of the world like brown rot *(Ralstonia solanacearum)* and ring rot *(Clavibacter michiganensis ssp. sepedonicus)*. More commonly spread is soft rot (*Erwinia spp.*) Apart from differing susceptibility between cultivars, their spread and severity are firmly related to mechanization. At present there are no truly effective means of controlling *Erwinia* attack besides practicing hygiene, minimizing mechanical damage to the tubers and quickly drying the crop after harvest.

Tubers may also be attacked by and be the source of infection for the daughter tubers for a number of fungi like black scurf *(Rhizoctonia solani)*, silver scurf *(Helminthosporium solani)*, gangrene *(Phoma spp)* and dry rot *(Fusarium spp.)*. They are tuber borne and some of them (also) soil borne.

Multiplication schemes have evolved and were integrated into certification schemes that ensure that diseases mentioned above are not present at all or only at levels sufficiently low not to endanger yield, quality and health of the crop to be harvested.

For the buyer or user of the seed it is important to be aware of the quality of the seed either by buying certified seed or producing seed with help of a certification scheme, under proper climatological conditions and with the necessary inputs at hand.

Physiological health of tubers has to be maintained by proper storage of the seed either by the producer or the buyer of the seed, depending on the time of delivery. Physiological health or vigour shortly can be defined as the ability of the seed to quickly produce a plant and on the crop level to reach full canopy closure as soon as possible. To preserve vigour during long term storage in general, seed has to be stored dry and cool. Before planting the seed may be pre-sprouted to accelerate crop emergence and canopy closure. In case that planting is necessary relatively shortly after harvesting the seed, ageing may be accelerated deliberately by higher storage temperature.

Producing, storing and even using healthy seed is a very complex matter. Many aspects with regard to physiology are not yet fully understood. On all aspects of seed quality shortly introduced here - and on many others - a lot of information may be found in many publications and (hand)books, *e.g.* Burton (1989); Beukema & Van der Zaag (1990); Harris (1992); Mulder & Turkensteen (2005); Struik & Wiersema (1999).

Soil preparation, planting and ridging

The soil should preferably be free from pests, diseases, and stones which can attack or damage the stems and foliage, the tubers or both. Weeds may reduce yield and cause problems during harvest. To maintain soil health, potato rotation should preferably be wide, growing potatoes not more often than once every 5 to 6 years. Narrow rotations increasingly reduce yield and quality due problems with soil borne fungi like *Rhizoctonia*, *Verticillium* and *Colletotrichum* and due to potato cyst nematode (*Globodera spp.*) (*i.e.* Scholte, 1989 and 1990; Scholte & s'Jacob, 1990). Not all soil related problems are fatal to potato production. Some problems can be controlled by the input of fungicides (Rhizoctonia) or nematicides. They are however costly solutions that are also putting burden on the environment. Once introduced, most pests and diseases cannot be eradicated, but recent developments have produced tools that enable some of them to be much more effectively managed: *e.g.* potato cyst nematode (Been *et al.*, 2006) and wire worms (Ester *et al.*, 2004).

All soil handling should be carried out under conditions that do not harm the structure in the upper and lower layers in the soil. Especially handling heavy soils under wet conditions may compact the soil causing clods in the ridges that may reduce tuber set and may cause damage during harvesting. Compaction may also hinder root growth which decreases the rooted soil volume and therefore the amount of soil water available to the crop. Potato is not so much a shallow rooting crop but the rooting system is sensitive to adverse conditions like compaction (Dalla Costa & MacKerron, 2000).

Planting depth depends on both the soil type and moisture levels. Moist heavy clay soils cannot be tilled as deep as sandy soils. Especially when harvesting mechanically on heavy soils, planting should not be too deep to avoid harvesting too much soil and clods. Generally planting should be deeper in dry soil and at higher soil temperatures (Beukema & Van der Zaag, 1990).

Preventing greening of tubers starts with seed bed preparation of enough depth (8-10 cm) to plant the seed tuber sufficiently deep into loose soil. When tubers are planted sufficiently deep at a row width of 75 cm sufficiently large ridges (both high and wide at the top) can be built to provide enough "room" for the daughter tubers without many of them becoming green. On light sandy soils there is risk of soil washing down due to heavy rain. Having more soil on the tubers in large ridges decreases the risk of tubers becoming exposed to light. The top of the ridges should be centered on top of the position of the tubers. Then a furrow can be made at planting which is followed by the ridging equipment. For some cultivars with very oblong shaped tubers a row width of 90 cm offers more security for preventing greening which may outweigh the decrease in yield due to later canopy closure. Large ridges however do not provide full protection against green tubers. Especially on heavier soils greening may occur due to the soil cracking when becoming dry, exposing tubers (temporarily) to light. Proper hilling and preventing cracks is also relevant for preventing attack by potato tuber moth (*Phtorimaea operculella*) (Mulder & Turkensteen, 2005).

Fertilization

Fertilization is crucial to both yield and quality of potato tubers. Historically much work was done on the input of nitrogen, potassium and phosphate, optimizing both yield and quality. In many countries local research has provided potato growers with accurate advices on fertilization with NPK. Mostly they are based on soil testing and anticipation for local yield levels. Generally a certain minimum level for phosphate is advised to secure yield, at sufficient level phosphate is not influencing tuber quality.

Nitrogen determines yield very much, but at high rates it negatively influences most quality characteristics like dry matter content, fry colour and after cooking blackening. Therefore, advisory systems especially for nitrogen, which are based on monitoring the crop nitrogen status have been developed (Haverkort & MacKerron, 2000). They are aiming to supply the crop with just enough nitrogen for growth and to prevent overdosing.

The amount of potassium advised is generally much higher than is required just for yield, since by high potassium level bruising and black spot susceptibility of the tubers are reduced. In the Netherlands recommendations on potassium fertilization to ware and starch potatoes are largely based on the work and recommendations of Prummel (1956, 1966, 1986). Prummel (1956) states that on heavy clay soil in spring 300 kg K_2O equaling to 375 kg chloride can be applied without danger for yield depression. This indicates that potatoes are not particularly sensitive to damage caused by chloride. Research of the same author (Prummel, 1966) in starch potatoes resulted in lower starch yield due to the reduction of dry matter content, not in lower fresh yield. Prummels' later research into applying potassium for a complete crop rotation before the potato crop resulted in lower yield after winter/spring application compared to autumn application (Figure 1). Although Prummel (1986) did not describe the yield difference as a toxic effect, this information has attributed to the perception by farmers that the potato crop is more sensitive to the application of chloride than it actually is. Prummel (1986) observed that increasing chloride application of winter/spring application opposed to autumn application rarely caused typical symptoms of chloride toxicity but did cause earlier senescence of the crop. This makes it likely that nitrate uptake was reduced due to competition by chloride. Unfortunately this was never substantiated by field research in the Netherlands.

According to this background, better use of potassium chloride may be made than was done so far. Chloride is as effective as potassium in reducing black spot susceptibility both directly as indirectly by reducing the dry matter content (Jacob, 1959; Van Loon & Houwing, 1989; Maier *et al.*, 1986; Mondy and Munshi, 1993). Therefore it seems more efficiently to apply a smaller amount of potassium chloride shortly before potato planting rather than applying a high dose in autumn with the purpose of loosing the chloride by leaching during winter. The only concern may be that the higher dose may have to accompanied by a an increase of nitrate fertilization to compensate for the competition by chloride. As already indicated there are systems for monitoring the nitrogen status of the crop. Using such a system can detect a possibly too low nitrogen status and prevent negative yield effects of spring application of potassium chloride.
Also for the other elements than NPK there are recommendations for soil levels. If these levels are not met in general, it is advised to increase soil levels with appropriate fertilizers. If soil levels are sufficiently high, due to (weather) conditions crops may show symptoms of deficiency of for instance magnesium or manganese. If timely spotted, the deficiency can be relieved by spraying these elements and severe yield depression will be prevented.

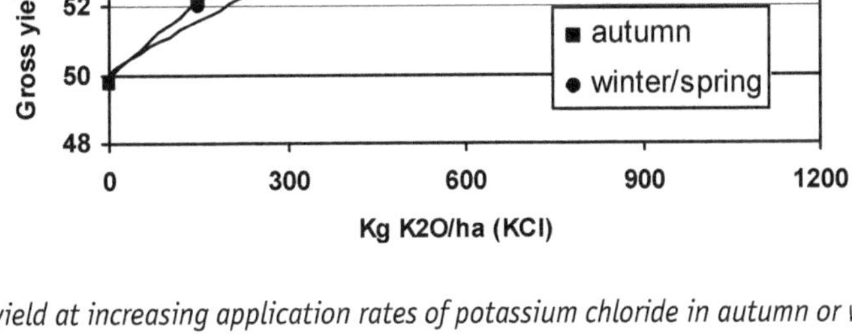

Figure 1. Potato yield at increasing application rates of potassium chloride in autumn or winter/spring before the potato crop on clay soil (data from Prummel (1986), modified).

Irrigation

The availability of irrigation for consistent yield and quality is essential in many parts of the world. Only few areas (consistently) have sufficient and timely natural precipitation. Even in The Netherlands that have one of the highest average yields, water supply is the most important yield limiting factor in potato production due to almost every year occurring periods of insufficient rainfall.

Moisture levels do however not influence the assimilation rate of the crop and therefore yield and tuber size, but during stolon and tuber formation it influences the number of stolons and tubers (and therefore tuber size) (Haverkort *et al.* 1990). Sufficient soil moisture levels during tuber formation and early growth prevents attack by common scab (*Streptomyces scabies*). Using irrigation to prevent standstills in growth is also important to prevent irregularities in the growth of the tubers which may cause or add to tuber defects like malformations, internal rust spot, growth cracks and hollow heart (Harris, 1992; Van Loon, 1981).

Crop protection and haulm killing

Since this paper deals with tuber quality and other papers in this book discuss crop protection, we will not treat diseases that primarily attack the canopy. We just state that adequate control is critical in that respect that the infection of both canopy and tubers may cause partial or even complete loss of the crop.

Haulm killing of ware potatoes at the end of the crop cycle is a useful tool to synchronize tuber skin set. Due to differences in senescence between parts of the field, between plants and even stems, tubers differ in growth rate and skin set. By ending the growth all tubers usually reach skin set 10 to 14 days after haulm killing. Unlike in seed potato crops, stems in ware crops mostly do not tend to regrow. Therefore under proper conditions, *i.e.* when it is not too wet and there is no serious infection of late blight, the haulm may be killed mechanically by flailing. It takes proper equipment: flailing machines are preferably front mounted, and good skills of the operator can prevent damages to ridges and tubers. Mechanical haulm killing fits well in a strategy of integrated potato production: not using chemicals when it is possible and when it is not endangering the crop (Vereijcken & Van Loon, 1991). Under wet conditions the haulm is usually killed by spraying. Advantage of spraying may be that weeds are killed as well.

Harvesting and store loading

The one main concern at harvesting and store loading is to prevent damages. Damaging the tubers may provide pathogens like soft rot, dry rot and gangrene with the opportunity to infect the tubers, potentially causing serious or even total loss of the crop. Damages, both external and internal also equals loss of quality influencing the acceptability and value of the crop as well as the quantity that remains to be sold.

To prevent damages it is important that planting and ridging was conducted properly. Next to this, a proper equipment is important, but even more important is to have personnel with knowledge and skills to operate the machinery. In order to minimize mechanical stress to the tubers, operation of the machinery can be optimized by using a so called electronic potato (van Canneyt *et al.*, 2003). With help of such a device the complete handling chain of potatoes may be analyzed as was done by Molema *et al.* (2000). With regard to mechanical stress it is important to be aware of the fact, that not only the severity of a single impact is important, but also the number of impacts even when they are relatively small (Molema *et al.*, 1997).

At store loading it is important to stack the potatoes to a height that allows the crop (considering cultivar, fertilization, length of storage period) to be stored without creating too much pressure bruising in the lower layer of tubers. Depending on the ventilation system pile height should not be too low either. It is also important not to store too much loose soil that will (locally) hinder ventilation and a good distribution of sprout suppressant.

Storage

To a large extent, storing potatoes is about maintaining both quantity and quality. Since potatoes often spend more time in storage than in the field, a lot of decisions have to be made and there is ample room for making mistakes. Since all cost of production and harvest and investments in storage have been made, successful storage is crucial to the economics of the potato crop. Storing potatoes is a specialism that requires a lot of knowledge on tuber physiology, tuber pathology, on storage equipment and software and on physics of air and moisture. Again, a lot of both background and practical information can be found in different books (Beukema and Van der Zaag, 1990; Burton, 1989; Harris, 1992; Rastovski and Van Es, 1987).

Primarily, when the tubers are harvested moist or even wet, drying of the crop immediately after store loading is most important, along with reducing the temperature when higher than 18 - 20 °C. At high temperatures infection and progress of different diseases may be very rapid, increasing losses and decreasing the quality of the tubers. Quickly drying reduces the rate of infection by soft rot, dry rot and silver scurf. Especially for silver scurf it is important that they remain dry as well. Rewetting of the tubers provides an opportunity for new infections by silver scurf, since this pathogen does not need physical damage for successful infection. Rewetting of tubers may occur by using ventilation air with a too high dew point, by having temperature differences within the store or by condensed water dripping from the structure or the roof of the building in times with outside temperatures that are much lower than inside the store. The first is an error that must be avoided, the latter two may - to a certain extent - be avoided by practicing sufficient internal ventilation. Sufficient external ventilation may be an important issue as well to keep carbon dioxide (CO_2) levels below values that negatively influence fry color by increasing the content of reducing sugars (Veerman, 1996) (Figure 2).

Temperature management should be adapted to cultivar and destination of the potatoes. Cultivars requiring a different temperature management should not be stored together but in separate stores. Especially with regard to levels of reducing sugars that are relevant for processing into French fries and crisps, temperature management is critical and demands a lot of monitoring, knowledge and skills.

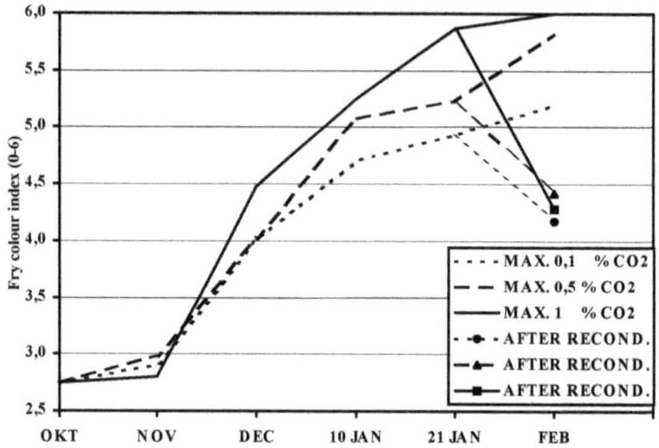

Figure 2. Effect of elevated CO_2 levels in storage on fry colour of French fries and effect of reconditioning (2 weeks 15 °C, in absence of CO_2) on increased fry colour index due to elevated CO_2 levels (1993/1994).

Discussion and conclusion

The production of a high yielding and high quality potato crop may fail in many ways. Due to the high cost of production, the concept of risk management is important in potato production. All major risks have to be manageable to make potato production an economically viable activity. When this is established, the optimizing process continues by always looking for the next weakest link, improving it and looking for the next weakest link. The improvement should be appropriate for the weakness: there is no point in fertilizing with GPS as long as there are serious shortcomings in fighting late blight.

It can be concluded that confidently starting potato production has a relatively high threshold: from the start a lot of production factors have to meet a minimum standard to sufficiently minimize failure of a high cost level crop.

References

Been, T.H., C.H. Schomaker & L.P.G. Molendijk, 2006. NemaDecide: a decision support system for the management of potato cyst nematodes, In: Haverkort, A.J. & Struik, P. (eds.): Potato in progress: science meets practice, Wageningen, Wageningen Acadamic Publishers, 143-155.

Beukema, H.P. & D.E. van der Zaag, 1990. Introduction to potato production, PUDOC, Wageningen, 208 pp.

Burton, W.G., 1989. The Potato, 3rd edition, Harlow, Longman Scientific & Technical, 742 pp.

Canneyt, T. van, E.Tijskens, H. Ramon, R. Verschoore and B. Sonck, 2003. Characterisation of a potato shaped intrumented device, Biosystems Engineering 86, nr. 3, 275-285.

Dalla Costa, L. and D.K.L. MacKerron, 2000. Plant and soil water status, In: Haverkort, A.J. & Mackerron, D.K.L. (eds.): Management of nitrogen and water in potato production, Wageningen, Wageningen Pers, 175 - 187.

Ester, A., K. van Rozen and F. Griepink, 2004. Tackling wireworms in a new way, International Pest Control 46, nr. 2, 80-81.

Harris, P.M., 1992. The Potato Crop, 2nd edition, London, Chapman & Hall, 909 pp.

Haverkort, A. J. and D.K.L. MacKerron, 2000. Management of nitrogen and water in potato production, Wageningen, Wageningen Pers, 353 pp.

Haverkort, A. J., M. v.d. Waart and K.B.A. Bodlaender, 1990. The effect of early drought stress on numbers of tubers and stolons of potato in controlled and field conditions. Potato Research 33: 89-96.

Jacob, W.C., 1959 Studies on internal black spot of potatoes, Memoir 368, Agricultural Experiment Station Cornell University, Ithaca, New York, 86 pp.

Loon, C.D. van, 1981. The effect of water stress on potato growth, development and yield, American Potato Journal 58, 51-69.

Loon, C.D. van and J.F. Houwing, 1989. Het effect van een chloorbemesting op blauwgevoeligheid, onderwatergewicht, bakkwaliteit en opbrengst van enkele consumptie-aardappelrassen, In: P. de Jonge (ed.), Jaarboek 1988/'89, Lelystad, PAGV, Publikatie 49, 241 pp.

Maier, N.A., A.P. Hahlenburg and A.B. Frensham, 1986. Potassium nutrition of irrigated potatoes in South Australia. 3. Effect on specific gravity, size and internal bruising of tubers, Australian Journal of Experimental Agriculture 26, 737-744.

Molema, G.J., B.R. Verwijs, J.V. van den Berg and H. Breteler, 1997. Effect of repetitive impacts on subcutaneous tissue discolouration in potato tubers, Netherlands Journal of Agricultural Science 45, nr. 1, 185-200.

Molema, G.J., A. Bouman, B.R. Verwijs, J.V. van den Berg and J.J. Klooster, 2000. Subcutaneous tissue discolouration in ware potatoes. 1. A chain analysis in the Netherlands, Potato research 43, nr. 3, 211-224.

Mondy, N.I. and C.B. Munshi, 1993. Effect of maturity and storage on ascorbic acid and tyrosine concentrations and enzymic discoloration of potatoes, Journal of Agriculture and Food Chemistry 41, 1868-1871.

Prummel, J., 1956. Kalibemesting bij aardappelen, Stikstof 11, p. 367-372.

Prummel, J., 1966. Schade door chloorhoudende meststoffen bij aardappelen, De Buffer 12, nr. 6, p. 128-133

Prummel, J. 1986. Invloed van kalium op de blauwgevoeligheid van aardappelen, Rapport 12-86, Instituut voor bodemvruchtbaarheid, Haren, 29 pp.

Mulder, A. and L.J. Turkensteen, 2005. Potato diseases, Den Haag, Aardappelwereld BV, 280 pp.

Rastovski, A. and A. van Es (eds.), 1987. Storage of potatoes, Wageningen, PUDOC, 453 pp.

Scholte, K., 1989. Effects of crop rotation and granular nematicides on the incidence of Verticillium dahliae Klb. And Colletotrichum coccodes (Wallr.) Hughes, in potato, Potato Research 32 nr. 4, 377-385

Scholte, K. and J.J. s'Jacob, 1990. Effect of crop rotation, cultivar and enmaticide on growth and yield of potato (Solanum tubersoum L.) in short rotations on a marine clay soil, Potato research 33, nr. 2, 191-200.

Struik, P.C. and S.G. Wiersema, 1999. Seed potato technology, Wageningen, Wageningen Pers, 383 pp.

Veerman, A., 1996, The effect of increased CO2 levels on fry colour during storage, In: P.C.Struik et al. (eds.), Abstracts of Conference Papers, Posters and Demonstrations of the 13th Triennial Conference of the European Association for Potato Research, Veldhoven, 14-19 July 1996, 419-420.

Veerman, A., 2001, Variatie in knolkwaliteit tussen en binnen partijen van consumptieaardappelrassen (Variation of tuber quality between and within lots of ware potato cultivars), Thesis Wageningen University with summary in English, ISBN 90-5808-482-5, 253 pp.

Veerman, A., P.C. Struik and C.D. van Loon, 2002. An analysis of the effects of cultivar, nitrogen, potassium, location and year on yield an quality of ware potatoes in the Netherlands, Abstracts of papers and posters, 15th Triennial Conference of the European Association for Potato Research, 14-19 July, Hamburg, Germany, 44.

Vereijcken, P. and C.D. van Loon, 1991. A strategy for integrated low-input potato production, Potato Research 34, 57-66.

Curriculum vitae – Arjan Veerman

Arjan Veerman studied arable crop science at the Agricultural University at Wageningen (The Netherlands) and graduated in 1988. In the same year he started his professional career at the Research station for arable crops and field grown vegetables (PAGV, presently known as Applied Plant Research) at Lelystad (The Netherlands). He obtained his doctorate at Wageningen University in 2001. His PhD thesis dealt with variation in potato tuber quality between and within consumption potato lots. His present research activities include work on potato storage disease control, sprout inhibition and quality management in storage. He has a major responsibility in potato research projects within Applied Plant Research. Furthermore he is active as advisor in several committees, one of these being the Advisory committee on seed potatoes of the Netherlands Inspection Service (NAK).

Curriculum vitae – Romke Wustman

Romke Wustman studied arable crop science at the Agricultural University at Wagenigen (The Netherlands) and graduated in 1976. In the same year he started to work for the International Potato Center (CIP) at Lima (Peru) and was stationed at Islamabad (Pakistan) working in Pakistan and Afghanistan. Subsequently he worked in potato projects in Bangladesh. He started to work for the Research institute for storage and processing of arable crops (IBVL, presently A&F-WUR) at Wageningen in 1984 with work in The Netherlands and assignments abroad. In 1990 he moved to the Research station for arable crops and field grown vegetables (PAGV, presently Applied Plant Research) at Lelystad (The Netherlands) and was in charge of the Netherlands potato varietal assessment programme. From 1998 till 2003 he analysed for the Netherlands starch potato processing company AVEBE the starch potato production at growers' level in The Netherlands and in Germany. His most recent work includes projects and activities within The Netherlands and abroad (central Europe and S.E. Asia).

Company profile – Applied Plant Research

Applied Plant Research (Praktijkonderzoek Plant en Omgeving B.V. (PPO)) is a modern enterprise which, in close cooperation with its clients, is working towards practical and sustainable solutions for agricultural crops, horticultural crops, mushrooms, trees and fruit crops. By means of combining forces and the integration with the Wageningen University & Research (WUR), Applied Plant Research is flexible with respect to research demands and is offering a broad scope of approaches. Our research facilities are located throughout the Netherlands, facilitating possibilities to conduct research under various situations. Within the organisation, practical skills are available on crop-, company-, and sector level as well as disciplinary level. PPO has four sectors for research of and demonstrations in:

- Arable crops, multifunctional agriculture and field grown vegetables (Lelystad).
- Bulbs and trees (Boskoop).
- Bees, fruit trees and mushrooms (Randwijk).
- Glasshouse horticulture (Naaldwijk).

Each sector has at least one research location. Besides specific knowledge of the crops (cultivation, quality control, integrated control of pests and diseases) and marketing; there is ample knowledge available on practical farm management.

Applied Plant Research boasts a broad network in the sector. Applied Plant Research holds a strong position within the scientific world, achieving that newly developed insights (foreign or national) can quickly be fitted into the research, where practical applicability plays the mayor role.

Applied Plant Research has obtained a central position within the primary sector and is very well known for its well attended demonstration days. Also the Applied Plant Research crop manuals, the contribution to and the organisation of courses and cooperation with farmer study groups enhances its strong position. Increasingly, Applied Plant Research has become more active abroad, including central and eastern Europe. Amongst its clients Applied Plant Research has the Ministry of Agriculture, Nature and Food Quality, The Commodity Board for Arable crops and the Commodity Board for Horticultural crops, Dutch Organisation for Agriculture and Horticulture cooperations and enterprises within the agricultural sector as producers of propagation material, crop protection products, agricultural machinery *etc.* Large internationals as Unilever and AVEBE have also been included in our extensive database of clients.

Link between knowledge, experience and capacity within Applied Plant Research.
Applied Plant Research boasts extensive knowledge in the areas:
- Practical solutions oriented research.
- Operating research farms.
- Knowledge exchange by means of demonstration days.
- Crop manuals of most crops cultivated in areas with a moderate climate.
- Courses and workshops.
- Project management.

The courses are presented internationally as well as nationally. As we are very well up to date concerning the potato, vegetables and fruit chain (AGF in Dutch), chain development and business development. Applied Plant Research Business Unit AGV has been mandated as programme coordinator for the EU Access Programme (central and eastern Europe), in which five new member states participate concerning the "market access of fruits and vegetables". Applied Plant Research AGV coordinates the activities deployed by these five new member states and those of the LEI (Agricultural Economical Institute) and Wageningen International Centre.

Applied Plant Research AGV also is involved in the establishment and execution of the DUCATT project in Ukraine and has initiated activities amongst others in Rumania, Sudan, Suriname, China, Vietnam, Kazachstan and Thailand.

Certification of ware potatoes and potatoes used for industrial processing in Poland

K. Zgórska
Plant Breeding and Acclimatization Institute Research Division, Jadwisin, 05-140 Serock, Poland

Abstract

The paper presents the current state of implementing the certification system in production of ware potatoes in Poland. The bases for the Polish Code of Good Agricultural Practice, integrated and ecological potato production and the rules of granting certificates have been described. The paper also includes examples of the HACCP system implementation in plants dealing with potato packaging and processing and EUREPGAP at large farms producing potatoes for industrial processing. The standard for potato merchantable quality applied in Poland is presented, too.

Keywords: integrated and ecological production, certification, EUREPGAP, HACCP, standardization

Introduction

For the last forty years immense changes in domestic and European agriculture have taken place. Intensive agricultural production is recognized as a factor responsible for soil and water pollution and constituting a threat to wild plant and animal species. Another important aspect is consumer search for food produced in conditions of limited chemicals in agriculture , food which is perceived as safe to health and characterized by high nutritional values. Nowadays, it is required that agricultural production means environment protection and production of quality food. That may be achieved by means of sustainable technologies which ensure sound management of soil resources restricting erosion and aiming at eliminating or reducing the amount of mainly pesticides and fertilizers used.

At present there are two main systems of sustainable agricultural production, being alternatives to conventional agriculture: integrated and ecological ones. The idea of Integrated Farming System (IFS) emerged in the 1980s in western Europe. The first research and implementation projects were initiated in The Netherlands, Germany and Switzerland. In integrated farming, the amount of pesticides and nitrogen fertilizers used has been substantially reduced (Kuś 2005).

The potato is still one of the main crop species in Poland although its acreage accounts for just 6% of the total cultivated area. Approximately 1.4 million households in the country deal with cultivation of that crop plant. That is rather unfavourable phenomenon in terms of market requirements. In most farms, the conventional system of potato production in Poland means a very low level of technology using chemicals, however, there is an increasing number of farms where use of pesticides and fertilisers is growing. The market potato offer from the two farm types is indistinguishable as it is not labelled with any marking or certificate, except for the marginal ecological production. (Nowacki 2006).

Cultivated potatoes can meet the requirements of plants originating from integrated production only if a producer is competent with respect to plant protection methods and modern agro-technical requirements and applies the principles of good farming practice. Another requirement of integrated production is its legitimisation and a possibility of supervising at any stage by the authorised bodies.

Polish Code of Good Agricultural Practice

The Polish Code of Good Agricultural Practice was developed in 1999 (Institute of Soil Science and Plant Cultivation 1999) based on the following legal regulations: Polish Acts and Decrees, EU Directives, International Conventions and the Helsinki Commission Recommendations. The Code is composed of four basic chapters:

- environment protection, farm organization and management;
- conservation of agricultural species;
- water protection;
- air protection.

The Polish Code of Good Agricultural Practice is similar to the set of principles contained in the GAP (Good Agricultural Practice), and at the same time the national regulations specify more precisely what is the most important to individual regions. In Poland GAP is the basis for potato production for the majority of potato packaging plants as well as plants processing potatoes for consumption purposes, which introduce the HACCP quality systems.

Principles of granting certificates on integrated potato production

The principles of integrated potato production in Poland were developed in 2005 and they have been approved by PIORIN (Plant Protection and Seed Inspection Service) (www.piorin.gov.pl).

The most important technological principles specified in the system include:

- soil requirements and crop rotation;
- fertilisation - basing on an analysis of nutrient content in soil;
- variety selection considering quarantine diseases (potato cyst nematode, potato wart disease);
- planting;
- plantation care;
- irrigation;
- plantation protection against diseases (basing of the chemical programme for plantation protection against late blight and alternariosis on the disease occurrence computer system - PIORIN);
- plantation protection against pests;
- preparing plantation for harvest and harvesting;
- harvest;
- storage.

A certificate issued for a producer confirming integrated production is an official confirmation that potatoes produced fully satisfy the above requirements. Pursuant to Ordinance No. 1836 of 26 July 2004 of the Minister of Agriculture and Rural Development supplementing the Plant Protection Act, such a certificate is issued by a Provincial Plant Protection Inspector appropriate for a given region.

The same ordinance sets out the detailed requirements to be met by a producer applying for the certificate, including:

- Complete training on integrated potato production confirmed by a certificate issued by an authorized institution.
- Keeping records of the whole production process in the "Integrated Production Log Book", in which a producer includes data on a farm and then detailed data on the whole cultivation process in appropriate tables.
- Throughout the production process, making it possible for PIORIN staff to supervise the whole process in line with the methodology (manual) required of integrated production and allowing for sampling potatoes for necessary testing analyses (*e.g.*, to check pesticide residues).
- Appling to PIORIN for:
  - joining the system of integrated production;
  - issuing an integrated production certificate.

The certificate is issued upon meeting all the above requirements and remains valid for 12 months only.

The presented system is compliant with the standards adopted in the European Union and is similar to the EUREPGAP and QS systems.

Implementation and certification in the EUREPGAP system

In Poland, the interest in the EUREPGAP certificate has become apparent in the two last years, mainly in fruit (apples) and vegetable production. Large potato processing plants (*e.g.*, Farm Frites Poland S.A.) have also taken interest in production of potatoes in line with the system. Most often these are plants with implemented integrated quality system based on the ISO9001, GMP, GHP, HACCP standards.

In order to ensure the very rigorous raw material quality and health safety standards, companies have started to implement the system "EUREPGAP Fruits and vegetables, version 2.0, Jan. 04". The system covers raw materials, technical equipment and individual stages of the cultivation process: preparing soil, planting, cultivation, harvesting and storage.

Certification is performed by international companies, *e.g.* Farm Frites-ECAS (European Certification Body for the Agricultural Sector). The EUREPGAP standard covers 14 strategic fields which are subject to detailed checking in the audit course. They include:

- traceability;
- keeping records and internal checks;
- varieties and seed potatoes;

- field history and field management;
- use of fertilisers;
- irrigation;
- plant protection;
- harvest;
- crop service;
- waste and pollution management, recycling and reuse;
- staff health, safety and wellness;
- environment;
- complaints.

The EUREPGAP standard is becoming popular among major potato producers in Poland as the growing number of food producers and large market chains want to co-operate exclusively with farmers for whom safety and product quality issues are the main element of the quality strategy.

Certification in organic production of potatoes

The ecological system means the manner of management with sustainable plant and animal production in a farm, based on materials of biological and mineral, non-processed biologically, origin, that is, agriculture without agrochemistry (Kuś 2005).

In Poland, the first ecological agriculture principles were specified in 1991 in the EEC Council Regulation 2092/91/EEC of 24 June 1991, on organic production of agricultural products and marking agricultural products and foodstuffs (Official Gazette No. L 198, 22 July 1991, as amended).

After Poland's accession to the European Union, the Act on Organic Farming of 20 April 2004 (Journal of Laws No. 93, item 898) was passed. The law specifies tasks and competences of the authorities and organisational entities in the field of organic farming. In Poland, the system of quality control and certification has been established based on Article 3 of the Act on Organic Farming and it is composed of the following:
- The Minister of Agricultural and Rural Development who authorizes certifying entities to perform checks, issue certificates of compliance for production with environment-friendly methods, in line with the PN-EN45011 standard.
- Agricultural and Food Quality Inspection supervising the entities issuing certificates in organic farming.
- Polish Centre for Testing and Certification granting and being responsible for accreditation of certifying entities.
- Advisory Centre for Agriculture in Radom - implementation of organic farming.

Ecological potato production is low in Poland, and in 2004 certificates were issued for approximately 20000 tones of potatoes. Farmers' interest in such cultures is growing. The complex study of the ecological system of potato production prepared at the Plant Breeding and Acclimatization Institute is helpful in introducing the system (Gruczek *et al.* 2005).

Certification in the HACCP system

In the 1980s, FAO/WHO recognised the HACCP system as an effective measure to ensure foodstuff health safety. In January 2002, the project on European Union Regulation 2000/0178(COD) was prepared, and since 2004 the regulation has replaced Directive 93/43/EEC. The documents promote the policy "from the field to the table", that is, the main aim is to ensure food safety throughout the whole food chain.

The system has been adopted in a number of plants in Poland manufacturing potato products and packaging ware potatoes. Those plants developed a plan for supervision of potato culture which includes:
- accepting seed potatoes and auxiliary materials;
- storing seed potatoes and auxiliary materials;
- preparing fields for cultivation and fertilization;
- preparing agricultural machines;
- planting potatoes;
- timetable for plant protection procedures, choosing pesticides, spraying doses and quality;
- assessment of mechanic weed control effectiveness;
- potato harvesting - harvesting time, air temperature during harvest, technical condition of machines, machine velocity, combine setting, height at unloading into a trailer;
- potato sorting - potato appearance;
- delivery to a storage facility - transport mean and container cleanliness.

At each of those stages the control type and qualities tested are specified as well as controlling methods, frequency, person performing the control and record site. Potato production is based upon the Good Agricultural Practice. The second stage of HACCP implementation includes approval of raw material and its packaging. Quality control at that stage covers:
- potato approval - assessment according to the Ware and Early Potato standard (Ordinance of the Minister of Agriculture and Rural Development - Journal of Laws No. 194, 29 October 2003), determining nitrate level, hazardous metal content and pesticide residues
- approval of auxiliary materials;
- potato storage - checking humidity and temperature, sanitary assessment of facility indoors, presence of pests;
- potato sorting - calibration correctness;
- potato brushing;
- storage condition - temperature, humidity, sprouting, sanitary assessment of facility indoors;
- potato washing - operation effectiveness, machine technical condition, washing effectiveness, water quality assessment;
- potato drying;
- qualitative selection - checking a product with respect to retaining attributes specified in a product description;
- weighing potatoes to individual containers - checking potato weight and quality in terms of a number of internal and external defects;
- packaging;

- labels - in line with the "product labelling and identification" guidelines;
- packing ready product into multipacks;
- storage of ready-to-use products;
- product distribution - quality and quantity of dispensed goods, transport mean cleanliness.

For all those stages also the control type, qualities examined, controlling methods and frequency, person or laboratory performing the control and a record site are specified.

In 2005, the International Organization for Standardization (ISO) published ISO 22000:2005 standard "Food safety management systems - for any organization in the food chain".

ISO 22000:2005 very clearly sets out the approach to the analysis of threats that may occur and to the obligation to consider stages that took place before and after the stage under analysis, taking into account external services or relationships with other food chain participants. Food safety management has to cover all the food chain, "from the field to the table" to be effective and the system traceability should be an important element which binds all the food chain links together.

The European Union regulations emphasize the necessity for controlling all the food chain by means of food safety management. Such an approach imposes the necessity for assigning Food Safety Objectives - FSO for individual production stages and specifying the Acceptable Level of Protection - ALOP for consumer wellness and health, while taking into account safety standards for the environment, animals and plants.

The use of the Good Agricultural Practice serves all those purposes, the integrated and ecological potato production system and other systems and certificates, such as EUREPGAP, SQ.

Standardization of early and ware potatoes in Poland

The Ordinance of the Minister of Agriculture and Rural Development on the detailed requirements on potato merchantable quality (Journal of Laws of 29 October 2003, No. 194, items 1899, 1900, 1901) has been in force since 2003. The ordinance pertains to preparation of early potatoes and ware potatoes for marketing and merchantable quality control.

The ordinance has been developed basing on the earlier Polish norms and UN/ECE standards: FFV-30 (early potatoes) and FFV-31 (ware potatoes).

The following terms have been defined:
- **Early potatoes** - potatoes harvested before they are completely mature, marketed immediately after their harvesting and whose skin can be easily removed without peeling.
- **Ware potatoes** - potatoes harvested completely matured, with suberized skin.

Specific requirements pertaining to potato merchantable quality are listed in Table 1.
All the entities purchasing ware potatoes and dealing with their packaging apply this standard; it is used as a point of reference for certification under the HACCP and SQ systems (Förschler 2005).

Table 1. Specific requirements pertaining to merchantable quality of early potatoes and ware potatoes

No.	Qualitative requirements	Tolerance %	
		Early potatoes	Ware potatoes
1.	Tuber size:		
	1. minimum diameter of round and round-oval tubers	transverse diameter 28 mm	transverse 35 mm
	2. minimum diameter of long tubers	transverse diameter 28 mm	transverse 30 mm, and longitudinal twice as long as transverse
2.	Maximum proportion of tubers:		
	1. greening	up to 1%	} total up to 2%
	2. internal defects	N/A	
	3. rotting	up to 1%	up to 1%
	4. mineral or organic contamination	up to 2%	up to 1%
	5. of diameter smaller that that specified for tubers mentioned at no. 1	up to 3% (however, not shorter than 15 mm)	up to 2% (however, not shorter than 28 mm)
	6. with common scab	N/A	up to 3%
	7. damaged	up to 2%	up to 3%
	8. shapeless	N/A	up to 3%
	9. immature	N/A	up to 2%
	10. of other varieties	up to 2%	up to 2%
3	Maximum total weight proportion of tubers specified at No.2. p. 1 and 3-5	up to 5%	N/A
4.	Maximum total weight proportion of tubers specified at No 2 p. 1-7	N/A	up to 6%
5.	Maximum total weight proportion of tubers specified at No. 2	N/A	up to 8%
6.	Tubers:		
	1. with sprouts over 3 mm	N/A	unacceptable
	2. excessive humidity on the surface	unacceptable	unacceptable
	3. with damage due to heat	unacceptable	unacceptable
	4. mouldy	unacceptable	unacceptable
	5. with frost damage	unacceptable	unacceptable
	6. contaminated with pesticides	unacceptable	unacceptable

References

EUREPGAP (2004). Inspection timing guideline fruit and vegetables. Version 2, January 2004.

European Commission (2005). Guidance document on the implementation of procedures based on the HACCP principles and on facilitation of the HACCP principles, and on facilitation of the HACCP principles in certain foodbuisness, Brussels.

Förschler A. (2005). QS - management of potato quality assurance, Kartoffelbau, 9, 10: 389-391.

Gruczek T., Nowacki W., Zarzyńska K. (2005). Ecological system of potato production Instruction, IHAR, Jadwisin: 1-34.

ISO 22000 (2005). Food safety management systems - Requirements for any organization in the food chain.

Journal of Laws no. 194, October 2003 tems 1899, 1900, 1901 - The quality requirements to merchantable quality of early and ware potatoes.

Kuś J. (2005). Perspective and development of different systems of agricultural production. Abstr. of papers. Conf. Perspective of production and trade of potato in Poland. Bonin: 5-7.

Nowacki W. (2006). Principles of integrated potato production. Ziemniak Polski 1: 19-22.

Official Journal L 198, 22/07/1991. Council Regulation (EEC) No 2092/91: Organic production of agricultural products and indications referring thereto on agricultural products and foodstuffs.

Polish Codex of Good Agricultural Practice (1999). Institute of Soil Science and Plant Cultivation, Instruction: 1-74.

www.piorin.gov.pl

Curriculum vitae – Kazimiera Zgórska

Kazimiera Zgórska studied foodtechnology at Warsaw Agricultural University from which she received a MSc in 1967. Then she worked in Potato Research Institute in Storage Department in Jadwisin which joined Plant Breeding and Acclimatization Institute and took its name, she works in the same place.

She received a PhD degree in 1978 based upon her work on factors affecting quality traits of potatoes for fresh consumption and processing. In 1990 she received the DSC degree at Cracov Agricultural University based upon her work on biological and ecological factors affecting the susceptibility of potato tubers after - wounding blackspots. Since 1993 professor Zgórska has been a Head of Potato Processing and Storage Department and since 1997 she has been working as a Visiting Professor at Koszalin Technical University. She led research program on the influence of growth and storage conditions on quality of ware and processing potatoes, postharvest treatment of potatoes safety of potatoes, standarization.

She is the author of over 100 scientific papers and hundreds of columns and conference papers. She frequently travels abroad and in the country. She is consultant for the agrofood industry. She is a member of the EAPR Council.

Company profile – Plant Breeding and Acclimatization Institute

The Plant Breeding and Acclimatization Institute is the largest Polish research centre in multi-disciplinary area of crop improvement, germplasm conservation and enhancement. The Research Centre in Jadwisin is a part of the Institute and has two departments: Potato Agronomy Department and Potato Storage and Processing Department.

The Department of Storage and Processing has 7 activities. The main area of activities of this department is as followed: storage characteristics of cultivars, suitability of cultivars for food industry, technique and technology of potato storage.

Quality improvement and market development of ware and seed potato in Russia

E.A. Simakov and B.V. Anisimov
All-Russian Potato Research Institute, 23, Lorch Street, Kraskovo, 140051, Moscow region, Russia

Abstract

The Russian Federation is the largest potato producer in the world after China. More than 10% of the world potato production is planted in Russia. Potato is one of the main foodstuff products in the country. In 2005 the total planted area of potato field in Russia was about 3.1 million hectares and the gross yield was 37.4 million tons. In this paper the authors have presented the information regarding potato production, structure of usage, market development and quality improvement of ware and seed potato in Russia. To know in detail the actual problems of potato products quality improvement in Russia is very important for the development of the Russian and in general of the international potato sector.

Keywords: production, usage, market, certification, standard

Potato production

Traditionally potato production is located practically in all parts of the country. The main regions of potato production are Central (11 million tonnes), Privolgskij (9.5 million tonnes) and Siberian (6.3 million tonnes) regions. Other significant regions where potato is grown are Ural (3.2 million tonnes), Southern (3.3 million tonnes), Northwestern (2.5 million tonnes) and Far-Eastern (1.6 million tonnes) regions. Table 1 shows the data of the area planted and potato production in the main Russian regions.

Table 1. The potato production in significant Russian regions in 2005.

Region	Planted area ('000 ha)	Production (million tonnes)
Russia, total	3088	37.4
Central region	941	12.0
Northwestern region	220	2.5
Southern region	320	3.3
Privolgskij region	796	9.5
Ural region	208	3.2
Siberian region	498	6.3
Far-Eastern region	132	1.6

According to the management forms during the last 10 years potato producers in Russia are generally divided into 3 main types: agricultural enterprises (AE), peasants farms (PF) and small private farms (SPF).

Different types of farms have their own features of areas under crops, gross yield, crop capacity and the technical level and efficiency of potato production. After the transition from a government regulated economy to a free market in Russia the major part of potato production is transferred to private sector private small family farms. Most of them have very small potato fields (less than 1 ha).

Analysis shows that nowadays in Russia small areas with potato production units prevail. Annual potato production in the SPF sector average 92% of total gross yield. The most potato produced by SPF is done manually using primitive technologies and with minimal mechanization. Average crop yield is not more than 11-12 t/ha. We observed that the present storage system and technologies limit producers' capabilities to prepare potato for the market. For this reason the potato producers have to spend significant efforts to sell the potato surpluses. The large agricultural enterprises' share in the total gross potato yield was reduced from 34% in 1990 to 8% 2005. Average crop yield in 2005 was 15.6 t/ha. Yield in Central and Ural regions was above the average (more than 17 t/ha), but in the Far-Eastern region it was below the average (11.7 t/ha) (Table 2).

Structure of usage

The structure of potato usage as well as potato production is highly fragmented. For a large proportion of the rural population in the main Russian regions, potatoes are a semi-subsistence crop with production targeted at farm household needs for home consumption and animal feed. Therefore the usage structure is mainly defined by both table potato demands and consumption as animal feed.

During the last years the consumption of the table potato was constant and estimated at a level of 120 kg per person. The total annual fresh table potato consumption amounts to 18 million

Table 2. Potato yield in Russia in 2005

Region	All producers t/ha	Agricultural enterprises t/ha
Russia, total (in average)	12.2	15.6
Central region	11.7	17.1
Northwestern region	11.5	14.6
Southern region	10.3	12.3
Privolgskij region	12.3	15.6
Ural region	15.5	17.9
Siberian region	12.7	13.4
Far-Eastern region	12.4	11.7

tonnes, *i.e.* about 50% of all potato production in Russia is used as a table potato. As animal feed is used about 6-7 millions tonnes of potato. Usually in years of abundant potato crop increase of potato usage in animal feed is found more than in dry years. Demand for seed potato for all types of farms is 9 millions tonnes and potato suitable for processing is 0.5 millions tonnes. It is expected that the proportion of potato production for processing will be increased as food processing industry will be more developed. So, taking into account the fact that the storage losses are still significant (about 2 millions tonnes), the total annual demand for potato in Russia is about 36-37 millions tonnes.

Potato market development

Nowadays Russia has comparatively low level of potato market development. The ware/fresh potato market is mainly formed under the influence of the small private farms (SPF). They are dominant suppliers on the ware/fresh potato market and don't depend on the seasonal changes of sales volume, *i.e.* they determine the price level during the whole sale period. The SPF share is about 70% of total volume (about 3-3.5 million tonnes annually) of traded potato. But taking into account that the largest share of the harvest is used for the producers' own consumption the marketable surplus or proportion traded is thus comparatively small and the level of potato traded is not more than 10% of total national production.

The distribution system is not developed either and this is a very important problem of the Russian potato market. Nowadays we can observe many different variations in the potato distribution system including farm sales, street trading, market places, as well as specialized greengroceries and supermarket. Since a few years the supermarkets with increasing market share are the important supplier of potatoes for citizens.

The greengroceries, market places and street trading continue to keep a significant part of their market share, but problems of product quality in this sector are becoming more and more important as consumers are looking for guarantee of quality. Usually quality control is higher in supermarkets. Supermarkets increasingly demand guarantees regarding product quality from their suppliers. In turn these suppliers lay down these demands to their producers, especially to those ones, which collaborate on the basis of long-term contracts. In the past the average consumers used to buy potatoes mostly from the market places, street trading and partially in the greengroceries. Nowadays they more and more buy their potatoes in the supermarkets. So, the Russian potato sector is little by little beginning to find the real opportunities to improve the products quality. Thus the production of table potato is mainly oriented to the local market. The Russian potatoes practically are not present at the international market. Only small amounts of potato were exported to the republics of central Asia in the last years. As a result today a participation of Russia on the international market is considerable less significant than in the past. Import of potato also is very modest (about 2% of total gross yield) and it does not noticeably influence the potato market.

The search for new opportunities for strengthening of integration processes in this sphere including the import and export is very important, in order to keep track of developments at the domestic and foreign potato market.

Quality of fresh table potato

Nowadays in Russia quality improvement of fresh unpeeled table potato is becoming more and more important. The basic requirements for quality of fresh table potato, which is sold for catering and retail market distribution systems, is defined by the State Standard of the Russian Federation. In the light of international trade development it was necessary to harmonize the Russian national standard of fresh table potato with the international standards of quality requirements. First of all they are FFV-30 and FFV-31 EEC UNO standards, concerning the selling and the goods quality control of early potato and fresh table potato respectively, which go for sale to international market.

It is expected that harmonization of national standards with international requirements will allow the use of produced potato more fully and to meet the needs in potato of the whole population. Furthermore the state price adjustment will stimulate the production of high quality products. It will give an opportunity for producers to improve the seed industry, to extend the range of goods, to transfer modern technologies, machinery and equipment for presale products preparation. Also it will allow consumers to use high quality potato products.

According to the new Russian standard the fresh table potato is divided depending on realization term in early table potato (which is sold till 1 of September) and late table potato (which is sold after 1 of September). There is the similar classification of fresh table potato in the international practice (FFV-30 and FFV-31 EEC UNO standards). So the classification of table potato is also uniform to the international classification. Depending on quality the fresh table potato is divided into the following classes:

- early table potato in 1st and 2nd classes;
- late table potato in extra-, 1st and 2nd classes.

The tuber appearance and size are the major characteristics of fresh table potato quality. Tubers should be fresh, clean, whole, dry, free from defects and without sprouts. According to standard's requirements there are restrictions for content of tubers with insignificant defects, which don't make worse the marketable state of table potato. They are light-green spots till 2 square centimeters, which can be removed by peeling; not serious tuber damage (cuts, pits, chaps, dents); potato scab, oosporosis and wireworm damages.

Standard's requirements concerning the tuber size of fresh table potato for sale have significant importance. The difference between minimum and maximum tubers diameter in the same packing should not exceed 20 mm for "Extra"-class and 30 mm for the 1st class. Good quality of fresh table potato is defined by the absence of mechanical damages (cuts, pits, chaps, dents) and other defects such as change of tuber shape, second growth, potato scab, rusty potato tuber blotch. Taking into account differences in Russian agroclimatic zones, where fresh table potato is cultivated, the standard supposes that the contents of tubers with defects should not be more than 2% of weight for 1st class and 5% for 2nd class of early table potato. For late potato this index is 2.0% for "extra"-class, 5.0% for 1st class and 10.0% of weight for 2nd class.

In Russia traditionally the main part of table potato is stored, prepared and sold in farm or enterprise where it was produced. Nets of packing services and wholesale centers are not sufficiently developed and they are grading only small amounts of potatoes. The requirements of grading and packing of fresh table potato for sale are harmonized with international ones. The "extra"-class potato should be washed and potato of 1st and 2nd classes should be cleaned with or without water. Extra-class and 1st class potato should be packed into tissue, polymeric or mesh packets. However according to the standard the potato of 1st and 2nd classes meant for the catering trade don't need to be packed. The packets with potato should be marked with indication of class ("extra"-, first, second). The washed potato must have the last date of expire indicated.

The Russian consumers are very conservative regarding potato. Traditionally they prefer well cooked potato with white flesh. However the preferences are changing lately. For example, a part of the consumers today approves of varieties with more firm flesh and also with yellow pulp and red peel.

A very important development is that in Russia the modern consumers are increasingly concerned about their health and in line with this they developed a new attitude towards food. A part of consumers creates specific demand for potato varieties, types, production methods, *etc.* For example, nowadays some consumers prefer low starch content potato varieties and potato that have been produced in a so called environment-friendly way on lower fertilizer inputs and chemicals (organic production). With an increase of this specific demand the new varieties, which differ in cooking quality, size, shape and taste appeared at the market. Many Russian potato varieties, which were bred during the last years, became very popular because they met the specific consumer's demand. It is expected this trend will continue. The size and type of packages have changed also. Thus, many shops often sell potato packed in smaller packs, such as 1.5, 2.5 and 5 kg. For consumer, plastic (polyethylene) foil, colored for protection against light and perforated for ventilation, is mostly used for the packing of potatoes. Greengrocers also sell the potatoes in loose paper bags. Hopefully these changes will give to Russian consumers a more broad assortment of high quality table potato.

Convenience products and processed potato

Since a few years convenience potato products are becoming increasingly popular and nowadays they are already well known by Russian consumers. This category consumer can find a wide range of products: washed potato, peeled uncooked potatoes, peeled and pre-cooled potatoes and deep frozen potato products.. These products are being sold in supermarkets and specialized greengroceries.

Washed potato was well appreciated by many Russian consumers after several years of consumption. At the same time other more simple potato products such as peeled uncooked potatoes is a rather new development. With these products the consumers always have an easy and quick meal. Since these products are in line with the needs and preferences of many groups of consumers, it can be predicted that the volume of this products will continue to grow and it will occupy a more and more important place at the Russian consumers market. It is expected that this trend in Russian food processing industry will continue in the future. The deep frozen potato products as well as crisps (respectively, chips), dry potato and many other complete meal potato, and starch also did not go through a positive development in Russia during recent years. Now these kinds of products are imported into Russia from European countries mainly.

The import of processing potato into Russia is rising continuously. In 1995 Russia imported 20,000 tonnes of potato products, but this amount increased to 150,000 tonnes in 2004 (Figure 1).

Nowadays the level of potato processing to food products is very low in Russia. A lot of potato processing enterprises stopped fully or partly. Thereby production of processed potato decreased in fivefold in 2004 in comparison with 1990 (Table 3). In fact all of 40 potato starch factories, which were built in the Russian Federation, don't produce potato starch anymore. Nor do factories, producing dry mashed potato and alcohol from potato, They use only 3-10% of their production capacity.

Taking into account the importance of this branch, the Potato processing development Program in Russia has proposed and submitted a research and development programma to to the present Government of the Russian Federation for consideration. Within the framework of this programme the future prospects and new real ways for the development of potato processing in Russia are defined for the period to 2010.

Seed potato quality improvement

The international potato seed trade increasingly develops. Therefore it is necessary to unify and harmonize the Russian seed potato classification system with the European one, which is adopted by EEC UNO for the European Union, and with other similar national systems, which are used in most western European countries. According to the Federal Law "About Breeding

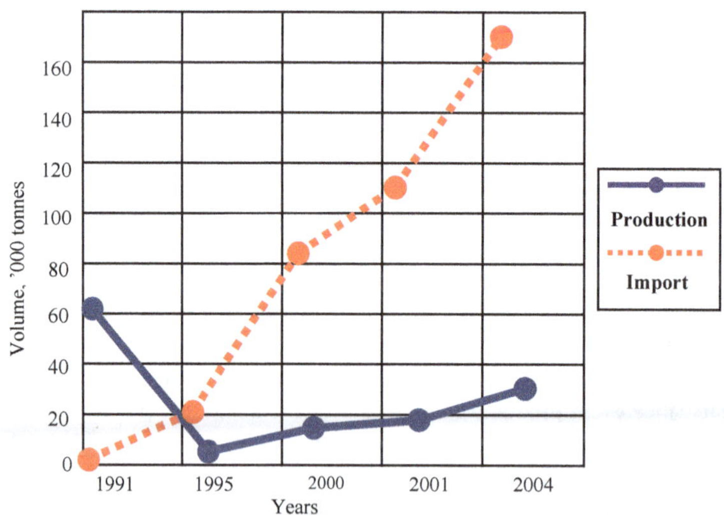

Figure 1. Changes of production of processed potato and import of potato products in 1991-2004.

Table 3. Processed potato volume in Russia 1990-2004 ('000 tonnes).

Product	1990	1995	2000	2001	2002	2003	2004	2004, % of 1990
Total processed potato products	133.3	10.1	12.6	15.0	14.9	28.8	27.8	21
including:								
dry potato starch	98.8	3.2	4.3	3.2	2.2	2.2	2.8	3
dried potato	7.9	1.5	0.4	0.9	1.6	3.3	2.1	26
quick-frozen potato	16.5	1.5	0.2	0.1	0.4	0.5	0.7	4
mashed potato and crackers	4.0	1.3	0.1	0.1	0.3	0.4	0.9	22
fried potato	6.1	2.6	7.6	10.7	10.4	22.4	21.3	35

and Seeds" in Russia seed potato is divided into 3 groups (categories): Original seed material (OS), Elite seed classes (ES) and Reproductive seed (RS).

OS is seed material which is produced by originator or his representative person. The originator is an organization or person, which created, bred, revealed and/or provides maintenance of a variety, listed in the official State list of breeding achievements, which are allowed to be used. Principal classes/generations of original seed potato can be represented by: Healthy mother potato plants selected in the field or derived on the basis of meristem culture; mini-tubers which were grown in the greenhouses, ionitoponic and hydroponic facilities; 1st field generation from mini-tubers and super-super-elite class. This generation has to correspond to high standard requirements regarding cultivar purity and minimally allowed levels of viral infections. In this respect the super-super-elite generation can be recognized as "S" Class (3-4 field generations) of European classification systems.

ES is seed material derived from original material and corresponds to state standard requirements and other normative documents of seed production. The number of elite seed generation is defined by the variety originator. Because of potato reproduction features, the numbers of elite seed generation can't be more than 2-3 field generations after super-super-elite. Therefore super-elite-I, super-elite-II and elite seed class are included in this group. These seed classes conform to "SE" and "E" classes of the classification system on the basis of field generation in most European countries.

RS are the next field generations from elite seed. For certification the inclusion of this category of 1st and 2nd reproductions after the elite seed class is most reasonable. There is no point in the use of more than the 1st and 2nd field generations after elite, because of progressive accumulation of viral and other diseases in the vegetatively multiplied potato. As a result the qualitative characteristics are degenerated and productivity drops with each next generation. Therefore the seed material of 3rd and following generations after elite can not be certified.

In conclusion, the division of Russian seed potato material into classes/generations is corresponding to analogical classification systems of most European countries. Differences are mostly in terminology of classes and categories names. The requirements of the quality of plantings and tubers that are used in Russia are shown in the Tables 4-5.

Table 4. Tolerance levels of seed crops in Russia (field inspection).

	Requirements for seed classes, max %		
	OS	**ES**	**RS**
Field inspection[a]			
Variety purity	100	100	97
Severe virus (PVY-mosaic and Leaf-roll)	0	0.5	1.5
Mild mosaic (PVX and PVM-mosaic)	1.5	3.0	9.0
Blackleg	0	0	0.5
Laboratory test by ELISA[b]			
PVY and PLRV	0.5	-	-
PVX, PVS, PVM	4.5	-	-

[a]*Phytophthora infestans* should not exist. The field must be free from Golden Nematode and Wart diseases
[b]Only for 1st field generation

Table 5. Tolerance levels ofr seed potato tubers in Russia (Lot inspection[a]).

	Requirements for seed classes, max %		
	OS	**ES**	**RS**
Variety mix	0	0	0.5
Black leg	0	0	0.5
Ringrot	0	0	0.5
Phytophora	0.5	0.5	2.0
Dry rot	0	0.5	1.0
Stem nematode	0	0	0.5
Scab (covered more than 1/4 of the tuber)	0.5	1.5	3.0
Rhizoctonia (covered more than 1/8 of the tuber)	0.5	1.0	2.5

[a]Seed potato tubers must be free from MOP-TOP and Rattle-viruses and from quarantine organisms. A Lot should not contain more than 3% by weight of tubers smaller the minimum size or more than 3% larger the maximum size indicated.

The main aspects of quality control are:
- variety control of original stocks. It includes 3-time field assessment during growing period; Elite classes - 2 assessments and reproductive classes - 1 assessment;
- all original seed material produced from micro-plants and mini-tubers are tested by ELISA-tests on the base of post harvest control.
- Lot inspections are realized on the basis of visual observations means of tuber analyses. The observations of super-super-elite, super-elite and elite classes are made by the accredited specialist of State seed inspection. Special certificates are given in accordance with results.

For quality improvement of seed potato in Russia the new project was started in 2004. A very important part of this project is the Bank of Healthy Potato Varieties (BHPV). This Bank is located and maintained in the Northern area (Archangelsk region) which has the cleanest phytosanitary conditions for viruses free material production. This Bank is a base for a new fund of original material free from viruses. The bank is used for Elite and certified seed classes production for all types of potato producers, including Agricultural enterprises, Peasant farms and Small private farms.

It was suggested that the seed production scheme be based on BHPV. This scheme allows to use the advantages of the northern phytohygiene with the complex application of modern methods and technologies of meristem culture and high-performance laboratory tests of virus detection (Table 6).

The most optimal prophylactic effect to be obtained in of seed potato cultivation the following requirements are met:
- Warranty of a complete absence of quarantine pathogens (potato wart disease - *Synchytrium endobioticum*, golden nematode of potato - *Globodera rostochiensis*, brown bacterial rot - *Ralstonia solanacearum*).
- Absence of soil viruses (TRV and PMTV) and their vectors (*Trichodorus spp., Spongospora subteranea*).
- Minimal risk of wide spread of phytopathogens in the field conditions (absence of PRLV, PVY, PVM virus pestholes and their vectors for a 2 km around).
- Minimal likelihood of bacteriosis spread (blackleg - *Erwinia spp.*, ring rot - *Clavibacter michiganensis*).

Table 6. Norms and methods of virus detection for different generations (seed classes) on the basic BHPV (Bank of Healthy Potato Varieties).

Generation (class)	**Testing norms**	**Methods**
BHPV and in vitro material	100% of plants	Elisa-test, NASH, PCR *etc.*
Minitubers in greenhouse	100% of plants	Elisa-test
1st field generation	200 plants per plot	Elisa-test
Super-super-elite	200 tubers (postharvest test)	Elisa-test
Super-elite	Visual symptoms assessment	Visual symptoms assessment
Elite and Reproductive classes	Visual symptoms assessment	Visual symptoms assessment

The possibility of viral infection of potato plants and virus transmission into tubers is determined by place and conditions of growing, level of infection pressure (amount of insect vectors and infection sources), resistance of cultivars and other factors.

Taking into consideration a wide diversity of agro-climatic conditions, infection level and number of insect vectors, we can mark out the three main groups of regions.

1. Northern, North-western and North-eastern regions have most favourable conditions for cultivation of quality seed potato. For example, Leningradskiy region is situated on the Baltic Sea coast. It has cool summer weather and relatively low background of insect vectors. These conditions conduct to minimum expansion of most heavy virus infections. The vegetative period in this region is very short - from end of May till middle of September (100-110 days), however this region has a very long light day (till 23 h). It creates very good conditions for fast potato growth and maturity. This agro-climatic conditions make this region favourable to cultivate quality seed potato.
2. Middle part of Russia (Central, Central black-earth and Volgovyatka regions, Ural, Siberia, Far East) has a wide diversity in the soil content and fertility, rainfalls, sum of active temperatures, frost-free period *etc.* However these regions can be characterized as regions with moderate infection pressure and can be considered as favourable areas to organize own quality seed potato production.
3. South and South-East regions have less favorable conditions to cultivate potato and organization of own quality seed potato production. These regions have a hot and dry climate during the vegetative period and a high infection pressure level (except the highlands in the North Caucasus area). Therefore the rates of accumulation of viral infections with each following field generation are much higher than in the Central regions. High rates of spread of infection decrease potato production. Quality of seed potato becomes much worse after 2-3 vegetative periods (or only after 1 vegetative period of susceptible varieties).

To ensure a transfer to the adaptation of a uniform seed potato certification scheme in Russia, the following activities are planned:

- Introduction of a compulsory set of requirements for virus control of original and elite seed potato categories, using laboratory test methods on the basis immune-enzyme analysis/ Elisa-test.
- The measures are necessary to develop a net of regional testing laboratories with modern equipment, technical means and diagnostic kits for a full range of laboratory tests for viruses, bacteriosis and other phytopathogens which are heavily decreasing the quality of seed potatoes.
- To organize the centers of phytosanitary monitoring in the regions on the base of research organizations with involvement of plant protection service specialists and regional State seed inspectors. Their activities should include the phytosanitary monitoring of movement dynamics of virus transferring insects, definition of their species and notification of seed production farms of the dates of mass aphid migration with the purpose of definition of the optimum dates for leaf tops removal on various seed potato plantings of various classes and categories.
- To arrange on the basis of the All-Russia Potato Research Institute and Certification Centre "Test Potato Service" the regular short-term trainings, practical training and consulting services for the specialists with the purpose to improve the practical skills in recognizing

variety properties and disease symptoms on potato plants and tubers, as well as training of specialists in post-harvest control of seed potatoes by means of Elisa-test.

Conclusion

Quality improvement of ware and seed potato facilitate the conditions for integration of Russia in the international potato sector.
The Ministry of Agriculture of the Russian Federation will provide the essential measures in order to improve quality of ware and seed potato. It is expected that the value to Russia will be an overall improvement of potato yields on farms to a level of 18-20 tonnes/ha, reduction of losses by 25-30% and the improvement of economic parameters of production. That will enable us to attract investments from various sources (Venture capital, entrepreneurs, interested foreign planters and other) for the development of the infrastructure, renovation and modernization of the storage base, improvement of technical and technological levels of potato production.

Curriculum vitae – Evgeniy A. Simakov

Evgeniy A. Simakov studied at the Agricultural University in Penza where he received a Master of Science degree in 1972. In 1977, after a short period of practical work, he was invited to the All-Russian Potato Research Institute, where in 1980 he received a PhD in Biology. In 1977-1985 Evgeniy Simakov worked in the Potato Research Institute as a researcher. In 1985-1996 he worked as a Head of Breeding Laboratory. In 1997-2004 - Head of Breeding and Seed Potato Center. Since 2004 until present time - Director of All-Russian Potato Research Institute. E. Simakov is a specialist in breeding and seed potato production.

His current research interests include problems of increasing genetic variability, breeding of new potato varieties and rapid multiplication of virus-free original material and basic seed potatoes. At present he is coordinator of the new program of potato research in Russia for the period 2006-2010. E. Simakov took part in national and international potato conferences and congress: Romania (1997), Finland (1999), Poland (2000), German (2002).

Curriculum vitae – Boris V. Anisimov

Boris V. Anisimov studied at the Agricultural University in Stavropol where he received a Master of Science degree in 1959. In 1966, after a short period of practical work, he was invited to the All-Russian Potato Research Institute, where in 1970 he received a PhD degree in Biology. In 1970-1980 Boris Anisimov worked in the Potato Research Institute as a Head of the Department of Seed Potato Research. In 1981-1986 he worked as a Director of Scientific Pathological Laboratory (SPL USSR) of the Ministry of Agriculture, Ambo, Ethiopia. In 1987-1992 he worked as a Head of Breeding and Seed Potato Center. In 1993-2003 - Head of Potato Section Program Development at the Ministry of Agriculture of the Russian Federation. Since 2003 until present time - Deputy Director of All-Russian Research Potato Institute.

Boris V. Anisimov was an initiator and coordinator of seed potato research production and seed certification program development in Russia, the breeding and seed potato program development at the Potato Research Institute, was a team-leader responsible for the breeding and seed potato program (1970-2003).

At the present time he is an initiator of the development of a new concept for seed production in Russia for the period 2005-2010. Boris Anisimov is an active member of Russian Academy of Natural Science since 2001. He took part in national and international potato congresses including the Netherlands Congress in 1995, the Congress in Finland in 1999 as a member of the Organizing Committee, and the Netherlands Congress in 2005.

Company profile – All-Russian Potato Research Institute (VNIIKH)

The All-Russian Potato Research Institute (VNIIKH) was founded in 1930 in order to improve yield and quality of potatoes in Russia. The main objectives of All-Russian Potato Research Institute are coordination of the national potato research and development programme; improvement of genetic variability and breeding of new varieties; production and delivering original seed potato of high biological quality; development and promotion of modern potato technologies (growing, harvesting, storage); national and international cooperation development. The Institute includes the following research departments: Genetics, Breeding, Biotechnology, Seed potato, Agrotechnics and Fertilizers, Plant protection, Technology and Mechanization, Storage and processing, Marketing and management, Scientific and Technical Information.

For 75 years the All-Russian Potato Research Institute has been responsible for inproving the quality of potatoes in the Russian Federation. The Institute's core business in the coordination of the national potato research programme, the improvement of existing potato varieties and breeding of new varieties. Other key tasks are the production of high quality starch potatoes and the development and promotion of modern technologies in the field of breeding, harvesting and storing of potatoes.

The main results of All-Russian Potato Research Institute are the following: The breeders of the Institute created more than 90 potato varieties, amongst which there are new high yielding disease-resistant and very popular varieties, such as Jukovskiy Ranniy, Udacha, Lukianovskiy and others. The Institute annually produces about 80,000 micro-plants *in vitro*, over 1,000,000 mini-tubers on which basis the Institute's experimental farms produce about 10,000 tonnes of original seed material (pre-basic seed material of the varieties recommended for growing in different regions). Up to 2,000 diagnostic kits for testing potatoes for viruses and bacterial pathogens are produces annually for the seed certification programme development. Fifty young specialists and farmers and 120 regional trainers annually attend the practical training course organized by Institute.

Trade of seed and ware potatoes

Reproduction right of seed potatoes in Russia

A.M. Malko
State Seed Inspection of Russian Federation, Orlikov side street 1/11, Moscow 107139, Russia

Abstract

The legislation basis of seed growing is the principal mover of its development. In recent years a considerable job has been done in Russia and no efforts were spared for the creation of laws and norms connected with seed growing and seed business. These documents take into account an international experience and expertise. The influence of that process is already visible in actual seed potato production. But it is worth mentioning here that there are certain problems to solve for further development of the potato seed market in Russia.

Keywords: law, legislation, seed production, breeder rights, certification, royalty

Introduction

From the moment of State establishment the seed law became an essential attribute. The regulative role of the State was expressed in a legislative basis followed by a strict control of its execution. Ten years of market-economy in Russia have shown an undeniable requirement of that kind of legislation for the seed business.

In the USSR the intellectual property rights for the plant varieties were not protected by the State norms or the law. All investments in breeding were from the State budget and the exclusive right for that property belonged to the State.

In the mid nineties of the last century a legislative basis was established reflecting worldwide experiences. The construction of legislative frames of the seed industry operation is the most important prerequisite of its successful adaptation to market based economy. The legislation of seed business and seed growing should be aimed to solve the following problems:

- providing the possibility of fundamental and applied research development in the sphere of seed growing and definition of the sources of financing;
- elaboration of the order, methods, and conditions of seed certification;
- regulation of interaction of the players of the seed market including export and import operations;
- identification of the forms of ownership for intellectual property and the system of protection of intellectual property rights;
- setting and adjusting the quarantine and custom regulation processes.

The mentioned notions were laid into the foundation of the national seed growing and seed business legislation. During recent years much work has been done to regulate the seed market of Russia. In order to fully recognize the national peculiarities of the seed industry, all parts of the laws and of the norms being elaborated were widely discussed and assessed at specialized meetings, professional seminars and scientific conferences. The experiences of the countries who achieve a great success in seed growing were taken into account. Important support and backing in elaboration of the modern legislative basis were given by international organizations in the frames of such projects as ARIS, TASIS, TRANSFORM.

To a certain extent the analogue legislative documents of the European Union and of Germany in particular served as an example for the creation of the national legislative basis of seed growing and seed business. However, it was impossible to transform the legislative basis from market-economy countries to Russia mechanically. The foreign documents were processed the way to be applicable to national conditions taking into consideration the seed growing traditions of Russia. For example, in Russian documents one will notice a more considerable role of the State in regulating the seed business and seed growing. Furthermore, tendencies for centralization of seed growing are more visible while market players have all possible freedom to operate in the market.

Present situation

At present, the policy in breeding and seed growing is built on two basic documents: the "Breeder Achievements Law" of 6 August, 1993, N5605-1 (Breeder rights protection law) and the "Federal Seed Law" of 17 December 1997, N149-F3) as well as on norms in the development of mentioned laws.

The first law ("Breeder Achievements Law" of the Russian Federation) regulates the interactions between creation, legislative protection and use of new varieties. The patented variety confirms the exclusive private right of the patent holder for its use. Any action with the seeds of the protected variety should be coordinated with the patent holder and confirmed by a license agreement at a bargain. The patent holder has the right of the collection of the breeder´s fee, which is called royalty. By means of that procedure, the money collected through the seed sale return has to finance the breeding process. In order to strengthen the application of the law a full packet of necessary documents was elaborated and approved.

The second and more important law for the industry (The Federal Seed Law) regulates production, harvesting, transportation, processing, storage, sales and other actions with seeds as well as the certification of seeds, and is the legislative basis of all these operations. The adoption of that law has allowed to order the interactions on the seed market and to set necessary "rules of the game". In order to develop the Federal Seed Law the packet of corresponding norms has been consistently elaborated and approved.

The most important traits characterizing the modern legislative basis of the industry are as follows:

1. The assertion that the Russian seed industry is already within market regulations has been proven. The variety has become an intellectual property and a full object of the market. The total complex of relationships between the State and producers, consumers and other players of the seed industry is transferred to a new quality level. The protection of rights for the variety represents now an exclusive right of patent holders. Any action with the seeds of protected variety should be confirmed by means of conclusion of the license agreement with the patent holder. Only the patent holder has the right of the breeder reward in the

form of a royalty. Collection of the royalty is considered to be a private right of the patent holder.
2. The exclusive right of a patent holder of the use of a new variety makes him fully responsible for the quality of seeds. It means that the quality of seeds should be high. The legislation provides to the State a possibility to define the frame conditions of the seed growing process and to control the fulfillment of the norms adopted by the industry. But at the same time this legislation has no right to reduce the tactical freedom of any participant of the seed market in choosing varieties, production technologies, equipment, marketing policies, sales policies *etc.* of whatever kind and nature as may be.
3. The modern legislative basis of seed business and seed growing with national peculiarities in general is unified with the legislation of other countries. It creates the real prerequisite for the integration of the Russian seed industry in the international seed market.
4. The rules of operations on the seed market of Russia are determined taking into consideration the regional peculiarities of Russia.

The process of improvement of legislation in question is under way. The Russian seed society now elaborates the amendments for the Federal Seed Law. The amendments appeared in the process of analyses of law applications from the previous period. All notions of the legislation in the sphere of seed business and seed growing are absolutely applicable to potato which is considered to be one of the most important agricultural crops of Russia. The total planted area of potatoes in Russia is more then 3 million ha. Being the leading producer in the world of potatoes for consumption Russia still has a very low potato yield per ha. Comparing with a worldwide average potato yield of 15 t/ha, the Russian Federation is in the range of 9-15 t/ha and the total production is about 35 million tons.

The low yield is due to several economic factors as well as to biological peculiarities of the crop. Within the technological process of potato production, breeding and seed growing take the leading role, because the vegetative reproduced potato tuber rapidly accumulates infection. One of the main factors determining the low level of potato yields in all Russian regions is the low quality of seed and its infection by pathogens.

The situation worsens due to the shortage of seed tuber production, grown in favorable conditions free of pathogen viruses, and due to the lack of technological regulations of the production of basic and reproduction seeds for planting purposes. As a result, the majority of potato fields is heavily affected by fungus, bacteria and virus infection. That is why one of the most vital factors of yield increase and efficiency of potato production is based on seed production at both federal and regional level. Thereby, the vast diversity of climatic conditions of Russia as well as the main influences on the seed tuber quality should be kept in mind.

Since 2001, legislation allows to certify the seed of varieties, which have been specified in a list of varieties admitted for cultivation and use in the Russian Federation (National Catalogue). The quantity of varieties in the list is constantly increasing. In 2006, 233 potato varieties were listed including 69 varieties protected by patents. The relatively low share of protected varieties in the list makes it more difficult for the breeders to collect the royalty. It turned out that approximately 66% of the 2005 planted potatoes were Russian varieties. The share of planting material from Germany, The Netherlands, and Belarus and that of other countries is also high

(Figure 1). Thus, varieties selected by Russian breeders, are the basis of variety resources. The share of foreign varieties has a constantly increasing tendency, indicating an integration of national and international markets. On the other hand, the lack of high quality seed from many varieties from the National List leads to numerous difficulties. An effective use of the genetic variability of the different varieties is not given For example, in 2005 the two varieties Nevsky and Udacha (Luck) comprised more than 50% of the total planted seed. At the same time, fourteen varieties represented more than 80% of all planted potatoes.

The annual demand of potato planting material in Russia is evaluated to 9 million tons of different potato categories. Actually the production of certified seed potatoes is not sufficient, but with assistance of the State Inspection Service the volume of certified planting material is increasing annually.

Conclusion

Future developments of the Russian seed potato market and raising competition abilities of the products need everyone's input on what is proposed below:

- unification with international requirements of production systems, quality control and certification of seed material;
- protection of intellectual property rights, widely spread introduction of field variety control into certification practice;
- simplification of the order of seed documentation excluding document duplication;
- formation of the All-Russian bank of healthy potato varieties, and creation of a fund of original seed material in the most favorable clean phytosanitary places of Russia.

To fulfill that tasks means to increase the quality of seed potatoes and to achieve a growth of potato production within the Russian Federation.

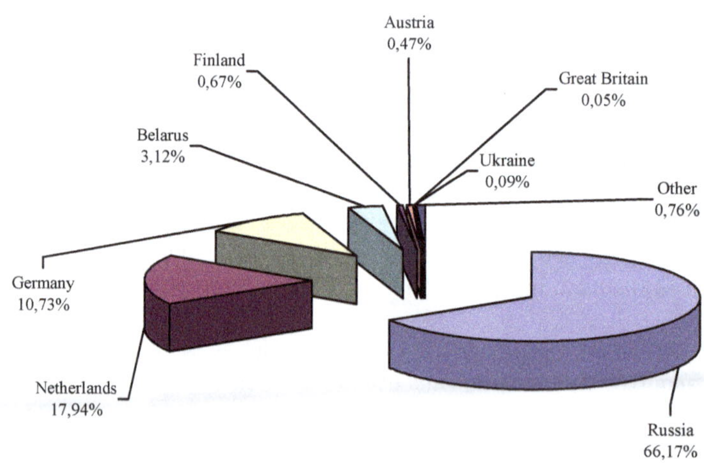

Figure 1. Shares of planted potato varieties in the Russian Federation from different countries (2005 data).

References

The list of varieties admitted for cultivation and use in the Russian Federation (National Catalogue) 2006. Vol.2. Moscow ExpressPrint Publishers: 5-9

The annual reports of Regional Seed Inspections of the Russian Federation, 2006

Curriculum vitae – Alexander M. Malko

Alexander M. Malko - was born in 1964 in the South of the European part of Russia. He studied agronomy at Timiriazev Agricultural University of Moscow, from which he received a MSc in 1987. He then worked for ten years for the seed breeding chair of the Timiriazev University. He received a PhD degree from the same University in 1992 based upon his work on methodology of research in seed breeding. At the same time he took part in elaboration of the Russian legislation of the seed industry, created of the basis of seed certification in Russia as well as in an educational process of the University. He is the author of over 120 scientific and educational articles and papers. He wrote and co-edited several scientific books. He frequently travels all over the Russian Federation as Chief of the State Seed Inspection of the Russian Federation, a position he took over in 2003.

Company profile – State Seed Inspection

The State Seed Inspection of the Russian Federation (Seed Inspection) is a Federal establishment with a headquarter in Moscow. It has 76 representative offices in 76 regions of the Russian Federation, covering all regions where plant production is present.
The Seed Inspection is a policy maker in performing the variety purity and seed quality control. It provides uniform scientific norms of seed quality and norms for the equipment for seed quality assessment for the whole territory of Russia. It fulfills methodological guidance, coordination and regulation of regional representative offices activities. The Seed Inspection creates and executes methodology of uniform application of the State standards and International rules while assessing variety purity and quality characteristics of seeds. The Seed Inspection organizes and fulfills the seed certification. It keeps the list of participants and objects of seed certification, and creates methods of seed tests and analyses. The Seed Inspection performs around 3 million analyses of seed samples. It provides approbation of about 25 million ha of crop production.

Risk Management: How to combine forwards with futures

J. Tietjen
HANSA Terminhandel GmbH, Schulstr. 2, D-27446 Farven, Germany

Abstract

The production, marketing and processing of potatoes has always been fraught with risks. There are many signs that their number and more importantly their significance will increase. Therefore risk management must be considered a central management task. Relevant risks are to be recognised and managed to ensure the existence and continued development of the company. In many cases it proves to be not only sensible to utilise several risk management instruments but at the same time to combine them. Thus, flexibility and usefulness can often be increased. The author demonstrates this using the example of EFP transactions: Exchange of Futures for Physicals. The marketing contracts combine the substantial elements of a forward contract in the physical market with those of hedging with futures at the exchange. Since the price determination is decoupled from the flow of goods, additional positive aspects arise. Overall, EFP transactions allow market participants to:

- simplify their contract negotiations;
- reduce their contracts' potential for conflict;
- take their quality standards into account in the contracts;
- eliminate their sale and/or purchase risk; as well as
- achieve price and thus calculation certainty.

Keywords: risk management, certainty, basis for calculation, marketing quality

Introduction

The occupation in agriculture but also in its up-stream sectors are ranked among those with the highest exposure to risk. This prominent role of risk is described by Moschini and Hennessy (2001) as follows: „(...) in agriculture they (uncertainty and risk, author's note) constitute an essential feature of the production environment (...).". Accordingly effective risk management is of high importance. Odening and Bokelmann (2001) define the former's task as limiting the negative effects of unfavourable environmental situations through effective measures and ensuring the important business goals of stability and certainty.

In the potato market, these goals can *e.g.* be affected by problems in the sale and/or purchase of goods. Consequently, their early organisation is in the interest of the market participants. They may utilise forward contracts for this (Bokelmann *et al.*, 2001). The contract negotiations are therefore often difficult or are abandoned without a result. Where the two contract partners lastly manage to come to an agreement, contract designs in many cases result in dissatisfaction with at least one party. Thus they often contain conflict potential from the start. Consequently, forward contracts constitute a possibility to manage the marketing/purchase risk but there is room for improvement with regard to the pricing. It seem sensible to decouple the price determination from the flow of goods. This approach can be realised with the EFP transactions - Exchange of Futures for Physicals - presented in this paper.

EFP transactions are marketing contracts that combine the main elements of forward contracts with those of price hedging at an exchange. In this case, the price for the goods is derived from the corresponding futures contract. Thus it always depends on the supply and demand situation at the time of delivery. Thus, beside the certainty of sale and purchase, positive effects result with regard to the contract relationship: There are no time consuming negotiations of a fixed price that develop disadvantageously for the buyer or the seller over time (ZMP 2001).

Beside the above positive aspects, the users profit from the effect of a price hedge. EFP transactions provide that at the time of delivery futures be exchanged for goods: Exchange of Futures for Physicals. Correspondingly, both parties to contract will use the opportunity to fix a price favourable for them at the exchange. They obtain price and thus planning certainty (Pflugfelder 1991). This in turn results in advantages for the financing of the business (Geveke 2003 and Fuhrmann 2001). Even other aspects of increasing importance - such as quality management - can be taken into account. In conclusion, EFP transactions can be described as a type of contract that is ideally suited in particular for potato merchants to carry out their quantity and quality management and at the same time ensure their profit margins.

Note 1. Potato futures of the Risk Management Exchange Hannover (RMX)
The RMX Hannover is an exchange under public law. Its competence lies in offering tailor made risk management solutions. In the potato sector, these are based on the trading of futures, of which the RMX Hannover offers three: the RMX Table Potato Future, the RMX European Processing Potato Future and the RMX London Potato Future. When trading potato futures, buyers and sellers must pay fees and post margin.

Note 2. Price hedging
The possibility for companies to fix prices for future purchases or sales of goods at an early date constitutes the core usefulness of price hedging. This works as follows: market participants intending to sell (buy) goods in the future, sell (buy) these early at the exchange and liquidate the contractual obligation initially entered into. Consequently, they always act in the physical and in the futures market (exchange). The goal of their transactions is the fullest possible compensation of losses arising in the cash market by value increases of the exchange position. In the ideal case, market participants can thus almost fully determine the economic result of their production or processing business even months in advance. The same applies to the future margin of trade transactions. In this way, the companies gain a solid basis for their calculations to which they can adapt their cash flows leading to an overall stabilisation of their economic results.

Sequence of an EFP Transaction

The sequence of an EFP transaction between a producer and a potato merchant is described below. The course of RMX Table Potato future price serves for better illustration. (see Figure 1)

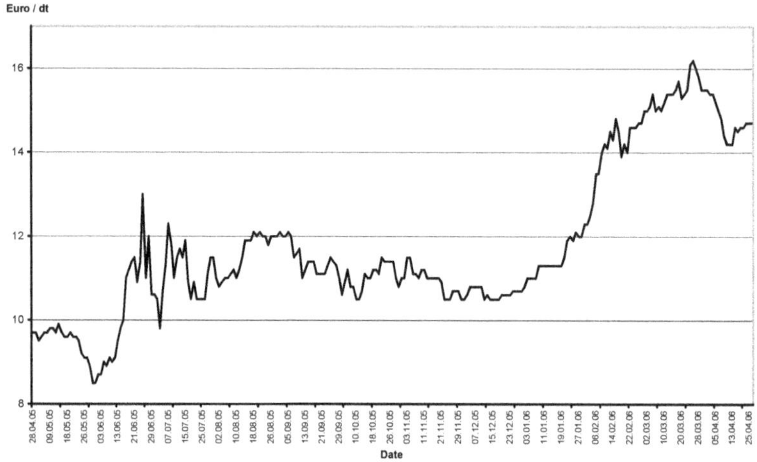

Figure 1. Course of the April 2006 RMX Table Potato future price (RMX Hannover 2006).

The example could be transferred to other supplier/buyer constellations within the value creation chain, *e.g.* trader/processor.

In May 2005, a potato merchant and a potato grower negotiate the delivery of 1,000 tonnes of table potatoes in spring of 2006. The potato merchant aims to achieve early supply certainty as planning basis for his contracts with the up-stream sector. Because of the voluminous harvest in 2004, the present price level in the physical market is very low. The potato merchant cannot profit from this level by entering a forward contract with a fixed price, as the producer expects rising prices. Since the two potential contract parties are in agreement about all other parts of the contract, they decide not to let the conclusion of the contract fail because of the differing price expectations but to carry out an EFP transaction. In accordance with the customary conditions, they conclude a purchase and delivery contract under private law, that contains the regulations with regard to all essential elements - quality, delivery location, delivery quantity, time of delivery (Tietjen 2001). For the price, they agree the price of the RMX Table Potato Future at delivery plus 50 Cents per 100 kg for a better quality condition. (also Becker and Bischoff 2002). As the goods are to be called continuously during the course of April, the parties to contract agree in addition to apply time bonuses. For those potatoes that are called before the last trading day of the RMX Table Potato Future (26 April 2006), the producer grants a discount of 10 Cents per 100 kg and week to the merchant.

Up to the time of delivery, both parties to contract have the opportunity to hedge a price that is favourable to them at the RMX Hannover. They can still react flexibly to market changes and at the same time profit from the already achieved sales and/or purchase certainty.

Because of the low price level, the potato merchant decides to hedge early at the RMX Hannover. During the course of June 2005, he buys bit by bit RMX Table Potato Futures until he has opened a total of 40 purchase positions. This corresponds to a quantity of 1,000 tonnes. On average, he achieves a price of 10 € per 100 kg. The producers postpones his involvement at the exchange. He expects rising prices and wants to participate in their future development. He starts only in mid-February to hedge his sales price. Until mid-March he sells RMX Table Potato Futures at an average price of 14 € per 100 kg.

Delivery of the potatoes takes place during the course of April. At the same time, the opened exchange positions are closed against each other. Based on the price of the RMX Table Potato Future and taking into account the time bonuses and the premium of 50 Cents per 100 kg for better quality conditions, an average sales price of 15 € per 100 kg results. Both must net this with the result of their price hedge. The merchant opened purchase positions 10 € per 100 kg that are now closed at 15 € per 100 kg. As the closing of a position occurs strictly speaking by entering into an opposite position- here: sale -, the merchant realises of profit of 5 € per 100 kg at the exchange. Thus he can realise a purchase price of 10 € per 100 kg. The producer also realises effectively the 14 € per 100 kg hedged in advance. He must subtract a loss of 1 € per 100 kg from the exchange transaction from the price of the potatoes (15 € per 100 kg). At last, the market participants must calculate an additional 25 Cents per 100 kg of transaction fees for the hedge.

Additionally it should be remarked that specific risks of futures trading can be excluded via EFP transactions. In particular the market depth risk and the basis risk are often cited as a reason for not using futures (Pennings and Meulenberg 1997). With the opening of the position these risks cease to exist for both parties. Correspondingly, the realisable price risk reduction does not have to be weighed against arising specific risks, as the case may be (Lampe 2003).

Note 3. Elimination of risks specific to the trading of futures

There is a close relationship between the cash market price of a commodity and the associated futures price: As a rule they converge up to the last trading day of the future up to a remaining difference. This price difference between cash and futures price is called the basis. For agricultural commodities the following count among the factors that influence the basis:

- differences in quality;
- cost of transport to the delivery location;
- cost of carry as well as;
- acquisition and handling costs.

The basis is specific for each region and each point in time. It may fluctuate over time and its value may be negative as well as positive. With respect to price hedging, it is important to know one's own basis. This is the only method for estimating the price that can be hedged for the company. Relevant for the effectiveness of a price hedge is the value of the basis at the time when the position is closed. If it corresponds to the expected basis that entered the calculation of the hedger, there will be a full compensation of losses from the cash market transaction from the appreciation of the position at the exchange and vice versa. If, however, the basis changes, the effective result will differ from the calculated one. This risk is called basis risk.

In case of a normal price hedge, the settlement price for the goods equals the cash market price. In case of EFP transactions, it is linked to the futures price by a premium or discount.

Accordingly, there is no more basis risk once the position has been opened as the settlement price and the closing price differ only by the agreed premium/discount.

Market depth risk is caused by disequilibrium of orders at an exchange that result in sudden price changes. While for relatively small sales (purchase) orders, the transaction price will correspond to the bid (offer) price, this is not the case for large orders in markets with low market depth. For large sales (purchase) orders, several transaction prices will result. This can be attributed to the fact that the principal will have to make price concessions for the complete settlement of his order. Depending on the order size and the number of market participants, the transaction prices for the sale (purchase) will thus decrease (increase) more and more to attract market participants for the opposite side. As a consequence of the market depth risk, futures markets with low market depth (*e.g.* relatively new markets) are less attractive for the market participants.

With the opening of the position, the market depth risk becomes irrelevant for EFP transactions: The positions of the two contract partners are swapped off-exchange. The futures price is consequently not affected. Even when transacting a large order neither of the two contract partners is subject to the risk of changing transaction prices (closing prices). The price at which the positions are swapped corresponds to the previous day's settlement price.

Conclusion

It was shown that EFP transactions are a promising type for the marketing of potatoes. The parties to contract profit from the advantages of a forward contract as well as of those of price hedging. In addition, the quality of their business relationship improves while at the same time they gain flexibility. Overall, users of EFP transactions may:

- simplify their contract negotiations;
- reduce their contracts' potential for conflict;
- take their quality standards into account;
- eliminate their sale and/or purchase risk; as well as
- achieve price and thus calculation certainty.

Because of these positive aspects, EFP transactions are suited to improve the co-operation within the potato value creation chain.

References

Becker, C. and Bischoff, H. [2002]: Verträge ohne feste Preise: Exchange of Futures for Physicals - Warentermingeschäfte strategisch nutzen (Contracts without set prices: Exchange of Futures for Physicals - Using Commodity Futures Strategically), in: Neue Landwirtschaft, H. 02/2002, p. 28-30.

Fuhrmann, R. [2001]: Hedging - ein Risikopuffer auch für die Bank (Hedging - a Risk Buffer also for the Bank), in: Neue Landwirtschaft, H. 07/2001, p. 22-24.

Geveke, J. [2003]: Rating - Neues Zusammenarbeiten (Rating - New Co-operation), in: Landwirtschaftsblatt Weser-Ems, H. 05/2003, p. 52-53.

Lampe, H. [2003]: Erfolgsdeterminanten und Entwicklungsmöglichkeiten des WTB-Schweinefutures unter besonderer Betrachtung seiner Nutzung als Risikomanagementmaßnahme (Factors of Success and Opportunities for Development of the WTB Hog Futures Contract under particular examination of its Use as a Risk Management Measure), Göttingen.

Moschini, G. and Hennessy, D. [2001]: Uncertainty, Risk Aversion and Risk Management for Agricultural Producers, in: B. Gardener et al. (Hrsg.): Handbook of Agricultural Economics, Bd. 1 a, Amsterdam, S. 87-153.

Odening, M. and Bokelmann, W. [2001]: Agrarmanagement: Landwirtschaft, Gartenbau (Agricultural Management, Agriculture, Horticulture), 2. Auflage, Stuttgart.

Pflugfelder, R. [1991]: Der Beitrag von Warenterminbörsen zur Informationsverbesserung und Risikoabsicherung bei Agrarprodukten (The Contribution of Commodity Exchanges to the Improvement of Information and Hedging of Risks for Agricultural Porducts), in: Agrarwirtschaft, Sonderheft 128, Hamburg.

Tietjen, J. [2001]: EFP`s am Beispiel Kartoffeln (EFPs illustrated with Potatoes), in: ZMP: Am Warenterminmarkt handeln - Beispiele, Begriffe, Zusammenhänge (Trading at the Commodity Futures Market - Examples, Terms, Context), 3. Auflage, Bonn, S. 41-43.

ZMP [2001]: Am Warenterminmarkt handeln - Beispiele, Begriffe, Zusammenhänge (Trading at the Commodity Futures Market - Examples, Terms, Context), 3. Auflage, Bonn.

Curriculum vitae – Joachim Tietjen

Joachim Tietjen concluded an apprenticeship as wholesale and export trader after graduating from the Agricultural Grammar School Bremervörde. Since 1978, he is proprietor and managing director of a potato wholesale trading firm that is operated by the family already in the third generation in the region of Rotenburg/Wümme in Lower Saxony. He is also managing partner of *HANSA* Terminhandel GmbH, which he founded in 1997.There he offers the conclusion of futures on agricultural products at the RMX Hannover in the segment Commodities. Thus he is familiar with the physical as well as the derivative trading of potatoes.

He is author of a weekly published comment on the potato market. He holds the presidency of the Association of the Potato Traders as well as that of the RMX Potato Working Group and the membership in the RMX Exchange Council on an honorary basis.

Joachim Tietjen is 50 years old, married and lives in Farven, the location of his company.

Company profile – *HANSA* Terminhandel GmbH

The *HANSA* Terminhandel GmbH was founded in 1997 with the aim of facilitating as financial services provider price hedging at the Hannover Commodity Exchange - the RMX Hannover.

As founding member of this exchange, *HANSA* Terminhandel GmbH is continuously and substantially involved in the total turnover at the exchange. Three trained brokers ensure the prompt execution of client orders in the daily trading at *HANSA* Terminhandel. In addition, a back office organises a smooth expedition of business daily information. Beside a daily information service, there are further regular publications such as the potato market comment which is published by fax, email, in the internet as well as in pertinent trade magazines. The extensive information offered on the company's web-site www.hansa-terminhandel.de is updated several times per day.

Consumer behaviour and marketing

How to think like consumers... and win!

H. Priestley
British Potato Council, 4300 Nash Court, John Smith Drive, Oxford Business Park South, Oxford, OX4 2RT, United Kingdom

Abstract

Consumer research shows that British potato consumption will decline sharply unless the industry succeeds in matching its products and promotions to emerging consumer shopping and eating trends, particularly the drivers of convenience and health. Understanding consumers is the key to being able to target industry responses in order to defend existing sales and achieve new ones. Through the British Potato Council, the British potato industry is now five years into an ongoing programme of consumer research. As a result, it has an improved understanding of how consumers shop by meal occasion and how perceptions of potatoes and rival carbohydrates differ by lifestage. This improved understanding is helping to direct cross-industry and individual business responses, including new product development, packaging improvements and promotional activity. Because it lacks a strong food tradition and emotional relationship with food, Britain is different from many mainland European markets. Nevertheless, findings from the British work to understand consumers will have relevance to other markets.

Keywords: consumers, consumer research, meal occasions, lifestages

Introduction

In potatoes we have a great product. It should have a great future. Unfortunately, we can't sit back and assume consumers see it that way. British consumers yearn for convenience and are increasingly concerned about health. Consumer research shows they perceive potatoes to be lacking in both areas. They're wrong! If only the solution was as simple as telling them that. Instead, the answer requires us to understand what motivates them and what shapes their opinions. We need to change their perceptions and make sure we have the right products, promoted in the right way to capture their imagination.

Over the last five years, the British Potato Council has invested heavily in gaining this understanding. It's been a major undertaking. For example, in total over 25,000 interviews and 100 focus groups have been conducted. And, it continues. For example, we have just commissioned work to get a better insight into meal repertoires - what meals people prepare on a regular basis, why they choose them and how we can encourage them to cook more meals containing potatoes.

Before I outline some of what we've learned, and what we are doing about it, I am conscious that I am addressing an international audience. So first, it is important to put our experiences into context. Some things we have learned would apply in many countries, but others would not. So it is important to understand what makes the British market different.

If you come from a country with a rich food culture, where fresh, local, seasonal food is carefully selected, lovingly prepared and then eaten as a shared experience, British shopper's buying habits may make no sense to you at all. In Britain today, mainstream weekday eating is about speed. The food is unlikely to be local or seasonal, it will be prepared as quickly as is humanly possible and will often be eaten individually. Even where there is a family in the household, they may eat at different times or, if they eat together, it may be on trays in front of the television.

Of meals 51% are now eaten alone, compared to just 15% 20 years ago (TNS, 2005), which may go some way to explaining the low priority given to cooking.

Another reason is time pressure. Eurostat data from 1999 showed that 23% (4 million) Britons routinely worked more than 48 hours a week, compared to less than 10% of the population throughout the rest of the EU. Tellingly, the greatest increases in working hours were being recorded amongst mothers with dependent children. Little surprise then that in 2002, the Henley Centre found that 45% of Britons agreed with the statement: "I'm so tired in the evening, I don't have the energy to do anything". Average weekday meal preparation time now stands at just 19 minutes, compared to an hour in the 1980s (TNS, 2005).

Our relationship with food is also more functional than emotional and is therefore closer to that in the USA, rather than most European countries. Inevitably there are exceptions. We have a growing section of the population termed 'foodies'. They do take great care sourcing food and preparing it. For them it is not just about three meals a day, it is their hobby. So they spend accordingly - both in terms of time and money, although for some this applies only to weekends. This is a high value niche, but it is low volume. They are very much a minority.

Recent evidence (FSA, 2005) does suggest a wider shift. It appears that more people are spending longer reading labels and considering where food comes from. 'Junk' food sales are falling and families may even be making greater efforts to enjoy some meals together. These indications are also being reported privately to the BPC by a number of companies, so it is possible that if I were to make this presentation in three year's time, it would be significantly different. However, at present, these represent tentative steps towards a respect for food. It remains the case that Britons spend less of their disposable income on food (under 10%) than any other Europeans.

As in other countries, multiple retailers have enjoyed huge growth and 92% of consumers now use them for the majority of their food shopping. This includes over 80% of fresh produce purchases now, compared to 23% in 1983 (FSA, 2005).

Where greater importance or interest is directed to food, then this generally means meals prepared elsewhere and eaten at home ('takeaways') or meals eaten outside of the home. British Consumers are Europe's biggest spenders on eating out, spending around 451 Euros per person each year, compared to 426 by Italians and 360 by the French (Datamonitor, 2005). Here again it is very important to understand the context. Because we lack a food tradition, we have few 'British' food outlets. Among takeaways we are perhaps famous for fish and chips, but actually chip shops are far outnumbered by Chinese, Indian, kebab, pizza and burger outlets. Similarly, restaurants reflect many food cultures, except our own. Table 1 indicates just how open we are to 'imported' food cultures and how this has now also spread back to retail sales.

All this has implications. On the one hand, in the home, the drive for convenience should favour all potatoes because of their versatility, and processed products because of their speed. However, our main competitors - rice and pasta - are arguably performing better.

Meanwhile our lack of a food tradition and openness to foreign cuisine is a major challenge because many of our 'imported' food cultures do not rely on potatoes as the carbohydrate of choice. For example, Chinese and Indian meals are served with rice or speciality breads. Substituting potatoes into these servings is an unrealistic prospect. Even in that bastion of 'Britishness', the pub, we now see food menus full of rice and pasta dishes, where traditionally you would have seen sausages and mash, pie and chips, beef and roast potatoes. Perhaps more worrying still is that many of these traditional 'British' dishes are perceived as being unhealthy (BPC, 2005).

The international nature of our foodservice market and its focus on sourcing uniform raw material at minimum cost, is an issue for potatoes. The British Potato Council is active in this area and has conducted detailed research on buying criteria amongst foodservice businesses. However, this market is so distinct from the in-home market that the two cannot usefully be considered together. This paper therefore concentrates on the in-home market - food that consumers buy and prepare for themselves.

The fresh market is perhaps where we face the greatest battle in the British market. I say this because, arguably, our processors are already quite successful at doing exactly what this presentation is about - understanding consumers. They are succeeding on convenience. For instance, take the phenomenal growth of pre-prepared frozen roast potatoes. By 2002, these accounted for 27.6% of the frozen potato market by value and 34.2% by volume (TNS, 2003). And now they are fighting back on health. For instance, boldly telling consumers that oven chips are a low fat food and promoting crisps with 70% less saturated fat. 2004 data from leading crisp producer, Walkers, showed 8.9% volume growth for these 'healthier' crisps, compared to a 1.9% decline for their 'normal' counterparts (TNS, 2005).

In contrast, the fresh potato market has been slower to understand consumer thinking and react to it. Partly this is explained by the fact that it is largely unbranded, with smaller operators and has traditionally focused more on trading than marketing. This is changing and the industry

Table 1. Retail sales of ethnic foods - million Euros. Population sizes are broadly comparable at c.60 million (Leatherhead, 2004).

	Oriental food	Indian Food	Mexican food
United Kingdom	800	895	405
France	196	14	86

response to BPC market research, some of which is summarised in this presentation, is gaining momentum.

Five background themes

Having set the context for our relationship with food, it is also important to consider wider food issues that affect potatoes, but which are not specific to them. There are many such issues and these have been explored in various phases of our research. In our latest work, five current themes were evident:

1. 'Help me to be healthy'. Health issues connected to food, such as salt intake and obesity, are now highly topical and have been well publicised. Consumers now recognise the problem, but are unsure of the solution and are cynical. With the health 'bandwagon' rolling, they are bombarded with conflicting health messages, reports, expert advice, media food scares, and wonder treatments. In potatoes we have an inherently healthy product, but it is a challenge to get this message noticed in such a cluttered environment.
2. 'Convenience redefined'. For some time, declining culinary skills and time pressure have put emphasis on convenience. Now we see this trend developing separate threads. 'Shortcuts' like stock cubes and pre-prepared vegetables allow people to make things faster. 'Enhancements' like stir-in sauces and marinades allow people to make things better (or at least to a level beyond their own skills or interests). 'Meal solutions' go the whole way and deliver an ovenable or microwaveable meal without the need to prepare any separate components. It is here that Chinese, Indian and other imported styles feature particularly strongly. Interestingly, these convenience options are associated with varying degrees of guilt and this also varies with age - younger generations having least guilt and little sense of cheating. To them, 'ready meals' are a routine feature of a busy lifestyle.
3. 'Inspire me'. Consumers are generally repetitive in their eating patterns and there is strong demand for inspiration (although this interest is not always followed through). One of our top retailers has enjoyed excellent coverage by recognising this and providing simple, exciting tips at every opportunity - on packaging and point of sale materials. Many of these are the simple tips that might once have passed from generation to generation as part of learning to cook. Linked to this demand for inspiration is the fact that 66% of purchasing decisions are now taken in store rather than pre-planned (BPC, 2003).
4. 'Polarisation'. Both shopping and cooking are becoming increasingly polarised. With shopping this means a move away from the middle ground towards the bottom of the price structure and the top. So we have a huge proliferation of bargain offers and also huge growth in sales of premium products. Interestingly, the same shoppers will often buy at both price points. For example, our largest supermarket chain, Tesco, reports that 64% of shoppers buy from both its 'value' and 'finest' ranges. Similarly, with cooking we have quick, convenient food, or at the other extreme, largely at weekends, 'craft' cooking - particularly by 'foodies' as described earlier. For potatoes, the challenge is to have price and product offers at both ends of the spectrum and make these available at the right time.
5. 'Male shoppers'. Men account for one third or 10 million of our 30 million principal shoppers. 15 years ago, that figure was less 4.75 million out of the total of 25 million - equivalent to 19% (IGD, 2002). Importantly, men behave differently to women when it comes to shopping and the whole food industry is seeking to get a better understanding of this.

What we know about how and why they shop

In addition to these general trends affecting all food types, we also know quite a lot about how people shop. Remember here that we are talking about people wandering the rows of a major supermarket because that's how the vast majority of groceries are sold.

We know that people do not shop for a list of staple store cupboard items from which they can construct a flexible variety of meals according to need. Instead, at the time of shopping, they know exactly what type of meal they are shopping for. In other words, they shop by meal occasion. And that occasion will determine what they buy, whether the meal is likely to include potatoes and, if so, in what format.

For the in-home market there are basically seven such meal occasions (excluding snacking). These are shown in Table 2, together with the proportion of meals accounted for by each occasion. In looking at this, it is important to remember that, contrary to what may be typical in some other countries, different family members will eat at different times. In particular, children are often fed before both parents have returned from work.

The detail is too great for this paper, but we also know the current market size for each occasion and what the future holds for it. Similarly, we know which potato servings fare best within each meal occasion (*e.g.* roast potatoes for a family meal... the great British Sunday roast! or oven chips for a simple children's meal). And we know which carbohydrate competitor succeeds when potatoes are not chosen.

Meal occasions have been well understood by many sectors and well exploited. However, potatoes have perhaps been slower to respond, particularly in the fresh market, where as mentioned earlier products are unbranded and marketing expertise is still developing. Additionally, with potatoes, it is perhaps also necessary to have a better understanding of consumer perceptions before the practical opportunities become evident. The latest phase of our research has addressed this. We now understand more about how people perceive potatoes and, crucially, we know that this changes with lifestage.

Table 2. In-home meal occasions (BPC, 2003).

	Basic description	**Proportion of meals**
Family meal	Family eat together	22%
Quick and simple adult meal	Routine adult meal	23%
Quick and simple children's meal	Routine children's meal	12%
Healthy meal	Part of diet or chosen for low fat or calories	26%
Special adult meal	Special meal or treat for couples	7%
Children's treat	Special meal for children	7%
Entertaining	Prepared to share with guests	4%

What we know about how they perceive potatoes

Put bluntly, older people love potatoes and eat them frequently (see Figure 1). Those in younger lifestages eat them far less often. There is evidence that, unless we do something about this, it is a habit that people will carry through life. If that is the case, and we do nothing, then our market faces massive decline. In fact, a crude straightline extrapolation would suggest over a 17% decline in volume by 2015 and nearly 28% by 2025.

Segmenting consumers by lifestage is not new. Many other sectors do it. It works and, for potatoes, it helps us because we can identify the key drivers for each lifestage (Table 3). Again, it is too much detail for this presentation, but we also know how different types of potato serving fare in different lifestages and perceptions of our rivals like rice and pasta.

By now you will have noticed that health is an issue that keeps cropping up. It is also an issue common across most European countries. In our studies, healthy meals were the largest single meal occasion and health is a key concern for a number of lifestages. So it is worth looking behind the headlines at what consumers really think about potatoes in the context of health.

Health and potatoes - a big misunderstanding

If health motivates a British consumer, then the chances are they'll buy rice rather than potatoes. Of course, this is a sweeping generalisation, but it is borne out by looking at what consumers think. As Table 4 shows, consumers think potatoes fare badly on health and, to a large extent, this is because they think they are high in fat and calories - at least in comparison to rival carbohydrates. This would be uncomfortable for our industry if it were true. But it's particularly painful because it is utterly incorrect.

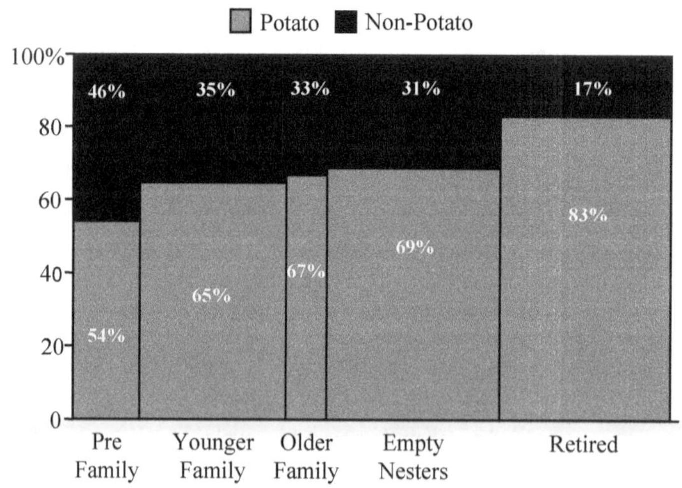

Figure 1. Proportion of meals served with potatoes according to lifestage. Width of columns denotes proportion of British population within that lifestage (BPC, 2005).

Table 3. Potential marketing approaches based on lifestages (BPC, 2005).

	Pre family	Young family	Older family	Empty nesters	Retired
Proportion of population	12%	24%	7%	29%	28%
Food drivers	Convenience, indulgence	Convenience, health	Convenience	Health, indulgence	Health
Offer focus	Frozen, chilled, exotic, different, fast	Fresh, chilled, frozen. healthy, nutritious, fast	Fresh, chilled, frozen, fast	Fresh, chilled, healthy, nutritious, exciting, exotic	Fresh, chilled, frozen, healthy, nutritious
Potential strategy	Fast and treat	Feel good	Fast and tasty	Quality	Fresh and healthy

Table 4. Perceptions (BPC, 2005) and reality (BNF, 2006) concerning the healthiness of potatoes.

	Potatoes	Rice	Pasta
Perceptions			
"Which is healthiest?"	27%	41%	25%
"Which is highest in fat?"	27%	9%	47%
"Which is highest in calories?"	41%	15%	33%
Facts (per 100g cooked weight)			
Fat (g)	0.1	1.3	0.5
Calories (kcal)	72	138	86

This misunderstanding is worsened by a press that delights in using chips and crisps as the foods of choice when attacking poor eating, obesity and a whole range of other social ills. Of course, they are less likely to report that by far the most common in-home serving of chips is oven chips. A 100g portion of these contains just 4.2g of fat (less than a small pot of natural yoghurt). It also provides a third of daily Vitamin C requirements and has more than twice as much fibre as a portion of brown rice. As mentioned earlier, crisp manufacturers have also made giant strides in reducing both fat and salt and, nutritionally, crisps compare well to products consumers may choose as substitutes, such as cereal bars.

Similarly, in our takeaway market, the traditional British serving of fish and chips is frequently attacked as being high in fat. But again, only half the story is told, because if someone who wants a takeaway does not choose fish and chips they'll eat an alternative takeaway. The fact is that an average portion of battered cod and chips contains at least a quarter of the fat and a third fewer calories than a serving of chicken tikka masala with pilau rice, or sweet and sour pork with egg-fried rice. It also has fewer calories, at least half the saturated fat and just a tenth of the salt of a cheese and tomato pizza.

Such is the importance of health as a driver, and so great are the misunderstandings, that the BPC has been directing a large proportion of its promotional budget to addressing this issue. Much remains to be done, and work is ongoing, but much has also been achieved. For example, we have campaigned on potatoes as a rich source of vitamins and minerals compared to rival carbohydrates and 53% of consumers now recognise potatoes as a good source of vitamin C (BPC, 2005).

Conclusion

My paper is entitled 'How to think like consumers... and win!'. I've outlined some of what we know about how consumers think. So I'd like to end by looking at how we can win! Unsurprisingly, I will argue that the solution is based on what we understand. As in any battle, winning is about defending what we have and attacking to gain more. And, as in any battle, focusing our resources is the key achieving real results.

By combining what we know about meal occasions with what we know about lifestages, we can do exactly that. Figure 2 shows this - taking each combination of meal occasion and lifestage and using data to show how well potatoes perform in that area. You will see the information is ranked according to that performance. So it shows where we are strong and should concentrate on defending by continuing to provide what consumers want. And it shows where we are weak and need to attack through product innovation and better promotion.

Our industry is highly integrated, with dedicated supply chains, so these can also overlay what they know about the specific customer demographics for the retailer they supply, thereby sharpening the focus still further. The challenge is great. For example, you'll see that among the areas we perform worst are healthy meals which is the largest meal occasion. However, the GB industry has already shown its ability to tackle the issues revealed by consumer research. For example, in our first round of research, household size was major issue. In fact, a third of homes had only one occupant. Yet our dominant packaging for fresh potatoes was 2.5kg bags. Worst still, our preferred promotion was 'buy one get one free' - so 5kg for someone only likely to use a fraction of the product.

Today the situation has greatly improved. 20% more volume now goes into smaller bags (up to 1kg) and there has been a corresponding fall in volume going into larger bags. Similarly, the early research emphasised how consumers wanted inspiration, simple tips and more useful information on bags. Again, there are many excellent examples of how our industry, fresh and processed, has made great progress in delivering exactly that.

We face many challenges. But because we know a lot more about consumers, and are aware that there is also more to learn, we are greatly increasing our ability to rise to these challenges and win!

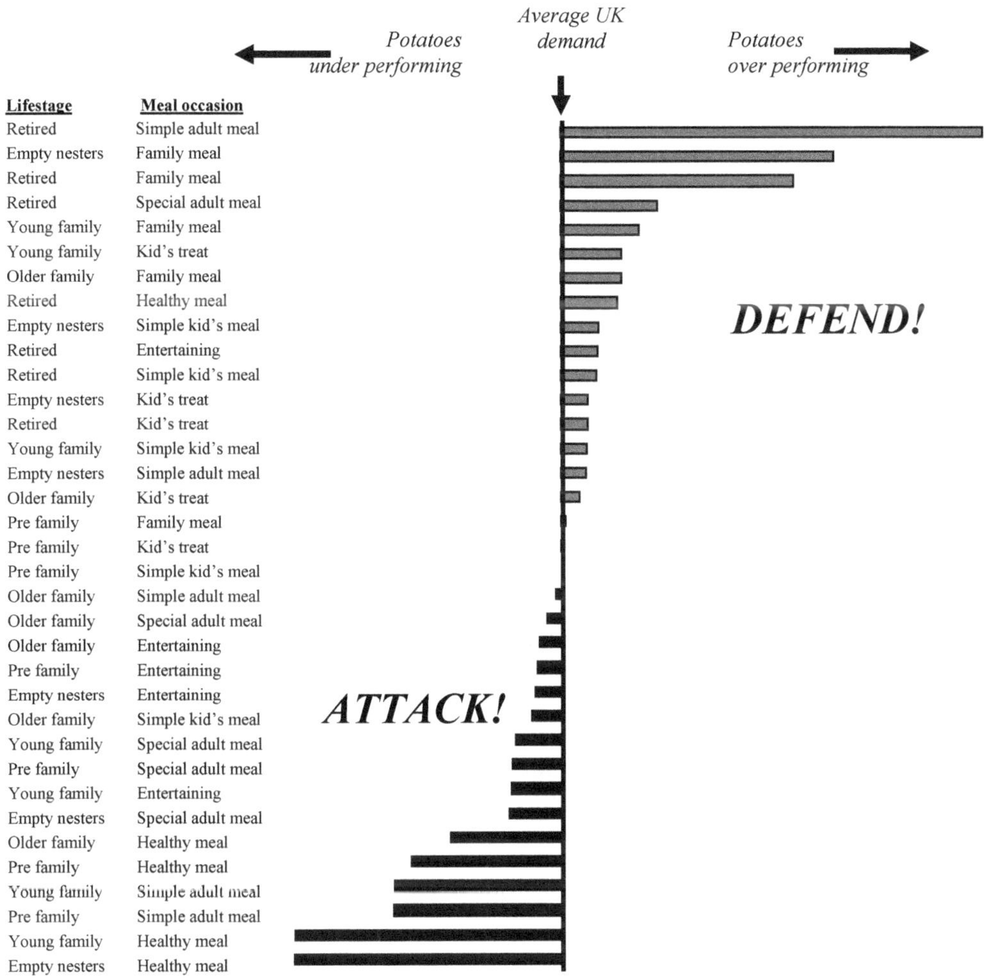

Figure 2. Potato servings per day by lifestage and meal occasion - where to defend and where to attack (BPC, 2005).

References

BPC, 2003. British Potato Council commissioned consumer research. Unpublished. Independently conducted according to the Market Research Society and British Market Research Association codes of practice. Predominantly desk research from existing TNS (see below) and BPC data.

BPC, 2005. British Potato Council commissioned consumer research. Unpublished. Independently conducted according to the Market Research Society and British Market Research Association codes of practice. Work included 1,200 face-to-face interviews at 40 locations, together with desk research and focus groups.

BNF, 2006. British Nutrition Foundation. Analysis prepared on commercial basis for the British Potato Council. Unpublished.

Datamonitor, 2005. Commercial, online, business intelligence service accessible by subscription at www.datamonitor.com
Eurostat, 1999. European Union statistics service accessible at http://epp.eurostat.ec.europa.eu
FSA, 2005. Food Standards Agency report: Consumer Attitudes to Food Standards, 2005. Annual consumer research. Downloadable at www.food.gov.uk/science/surveys/foodsafety-nutrition-diet/cas2005
Henley Centre, 2002. Commercial business intelligence service accessible by subscription at www.henleycentre.com
IGD, 2002. Institute of Grocery Distribution consumer research. Searchable free at www.igd.com Report: The single male shopper - are there men behaving badly?
Leatherhead, 2004. Leatherhead Food International report: The European Ethnic Foods Market 2004 (2nd edition).
TNS, 2005. Taylor Nelson Sofres. Commercial business intelligence service accessible by subscription.

Curriculum vitae – Helen Priestley

Helen Priestley - Chief Executive of the British Potato Council (BPC) since 2002. Previously BPC Director of Marketing and Communications. Joined the organisation in 1999 having spent 10 years in consumer marketing and the foodservice industry.

Completed a graduate programme at Forte Hotels, before moving to a key account role with Brakes, the UK's leading foodservice distributor. Then spent four years at Capital Radio as head of marketing for BRMB, Birmingham. During this time, research and innovative marketing helped keep the station at number one in an increasingly competitive market.

An honours graduate from Oxford Brookes University, she also holds a postgraduate diploma in marketing and is a member of the Chartered Institute of Marketing.

Company profile – British Potato Council (BPC)

British Potato Council (BPC) - Funded by the British potato industry via a statutory levy and works on its behalf to increase usage of potatoes and improve the British industry's competitiveness in meeting that demand. Active in the areas of consumer and export promotion, research and development, knowledge transfer and market information.

The British potato market is estimated at approaching £4 billion at the consumer level and £588 million at the farmgate. By volume, the market is split 50/50 fresh versus processed and also 50/50 in-home versus out-of-home consumption. The British industry is also recognised as one of the world's leading players, exporting seed and table potatoes, new varieties, science, consultancy and pre and post harvest technology all over the globe.

McDonald's Agricultural Assurance Programme

Dell Thornley
European Quality Director, Potato Products and Frying Oil, McDonald's Europe, Lyoner Strasse 34, D-60528 Frankfurt am Main, Germany

Keywords: McDonald's Agricultural Assurance Programme (MAAP), farm level

McDonald's Agricultural Assurance Programme (MAAP) provides a framework to manage sustainable and ethical agricultural production methods, as well as food safety and quality, at the farm level. MAAP's general policies lay out an overall direction, while product specific standards define details for production of McDonald's products.

In many European countries, local organizations or groups of farmers have developed local or regional quality programs. Rather than trying to create an independent quality label, MAAP is used as an internal reference against which these already-existing quality schemes are compared. The MAAP specifications then refer to the elements in existing schemes. If gaps exist between the specifications of existing quality programs and those in MAAP, McDonald's Europe initiates dialogue and develops a plan with program managers to close the gaps. By building upon the quality schemes already in place, we are able to make faster progress. To date, standards have been developed for beef, chicken, dairy, potatoes, wheat and lettuce. Under the auspices of SAI (Sustainable Agriculture Initiative) Platform's Working Group on Potatoes and Vegetables, further principles and practices for the sustainable production of crops are being developed and tested in on farm pilot projects. MAAP's future goals already comply with a high degree with SAI Platform's Principles and Practices for sustainable production of potatoes. McDonald's is the world's largest user and McCain Foods the world's largest manufacturer of French Fries. The companies are jointly working on two major initiatives relating to sustainability: McDonald's Agricultural Assurance Program (MAAP) and McDonald's Socially Responsible Supply Program, which at present is mostly dealing with environmentally-friendly processing.

Building on the learnings of MAAP, we are now working to expand our efforts to implement our environmental guidelines. Our starting point has been to collaborate with Conservation International and five of our direct suppliers in commodity areas important to our business (beef, bakery products, chicken, pork and potatoes) to develop and pilot-test a system for environmental goal-setting and continuous improvement at the stage of the food supply chain that they oversee - final processing. This system focuses on the four key environmental guidelines - water, energy, waste and air that are most relevant at the processing level and consists of a self-assessment and an environmental impact scorecard.

- *Water*. Water is a fundamental input to agriculture, food processing, and transport. Agriculture and food production systems should reduce withdrawal of ground and surface water by maximising water use efficiencies and eliminate the release of wastes and by-products into the environment via water.
- *Air*. Clean air is vital to the health of all living things. Agriculture and food processing industries can have both positive and negative impacts on air quality. Agricultural

systems should use practices that minimise the release of harmful by-products into the atmosphere.
- *Energy*. Although sunlight constitutes the most important energy requirement of agricultural systems, other energy sources are required for production, irrigation, harvest, transport, and processing of agricultural and food products. Production systems should maximise energy use efficiency and use ecologically sustainable renewable sources wherever possible.
- *Waste*. Most agricultural and food production systems produce solid waste. Waste increases production costs, leads to increased demand for renewable inputs that can exceed their rate of replenishment, and creates environmental problems if not treated and disposed of properly. Food and agricultural systems should minimise waste production, maximise recycling, and ensure proper handling and disposal of solid waste.
- *Soil*. Soil is the living foundation on which the food and agriculture industry is built. Soil maintenance and regeneration are vital to long-term agricultural production. Agricultural systems should maintain soil health by controlling erosion and improving the soil's structure and fertility to ensure stability and productivity.
- *Biodiversity*. Biodiversity is the sum of all life on Earth. Through complex interrelationships of species and their habitats, biodiversity provides vital ecological services such as pollination, flood control, climate regulation, and pest management. Food and agricultural production systems should protect native species and biodiversity by preserving natural habitats.
- *Pest Management*. Pests and disease pose constant threats to agriculture and food production. Agriculture systems should manage these threats to optimise production over the long term while minimising the use of chemicals and other inputs that impact human health, wildlife, and environmental quality.

The self-assessment is designed to provide a flexible format that allows suppliers to understand, monitor and communicate their social responsibility efforts and environmental performance. The self-assessment also allows suppliers to record and evaluate their social responsibility management practices, including policies, action plans, integration into strategic planning, training and participation in scientific research and stakeholder dialogue.

The scorecards are product-specific tools that consist of several performance indicators in each relevant guideline area. Rather than measuring adoption of specific food processing practices, the indicators aim to measure, wherever possible, actual impacts in each area.

Developing a global system for integrating our guidelines presents significant challenges that we will work to overcome in collaboration with suppliers and other external experts in agriculture and food production. These challenges include:
- Developing a system that is applicable to a wide range of industries.
- Accounting fairly for differences due to geography, scale and suppliers' experience with social responsibility issues.
- Developing a cost-effective yet functional approach to data collection and validation.
- Establishing appropriate goals and targets that are based on the latest scientific information and are able to stimulate continuous improvement.

We are already beginning to work on approaches to these issues. For example, in order to accommodate the variation in conditions around the world and to encourage continuous improvement by all suppliers, we plan to use the scorecard to work with each supplier individually to agree upon goals for performance improvement based on baseline performance results. Attainment of those goals will then be considered a favorable factor within our primary supplier performance assessment tool, the Supplier Quality Index, and we will work with suppliers to develop improvement plans when goals are not met. We expect the process to generate results that can be aggregated and analyzed to generate an assessment of McDonald's overall supply chain performance and to identify priority areas for action.

The challenges mentioned above are particularly pronounced at the farm and primary processing levels. Most of McDonald's direct suppliers do not control these stages of food production. As a result, we will need to identify innovative ways to encourage progress toward our guidelines at the farm level.

Curriculum vitae - Dell Thornley

Dell Thornley was born and raised in South Eastern Idaho. Dell graduated from Idaho State University Crops and Soil Technology in 1981. He then worked for Stukenholtz Consulting an agricultural consulting firm where he worked with growers on fertility programs and irrigation scheduling. He then headed to Saudi Arabia to turn the desert green; he spent 3 years there as Farming Operations Manager. He then worked for the J.R. Simplot Co. in Turkey, processing McDonald's french fries. In 1989 he joined McDonald's Quality Assurance Europe.

In 1994 Dell received the McDonald's International Presidents Award. In 2001 he led the Technical Advisory Group for the McDonald's produce suppliers which established McDonald's Good Agricultural Practices that later became the basis for the McDonald's Agricultural Assurance Program (MAAP) Standards.

Dell also works outside the McDonald's system on the Potatoes and Vegetable Working Group at Sustainable Agriculture Initiative (SAI) Platform.

Company profile - McDonald's Europe

Ray Kroc's first McDonald's restaurant opened its doors to customers in the United States in 1955. Today McDonald's is the world's leading foodservice company, with more than 30,000 restaurants serving over 47 million customers daily in 119 countries. McDonald's first restaurant in Europe was opened in The Netherlands in 1971. McDonald's Europe and its franchisees now employ over 260,000 people serving over 10 million customers every day in 42 European countries.

With revenues of USD 5.8 billion, McDonald's Europe contributes about a third to McDonald's worldwide revenues. Our European structure and this report cover all the European countries - from Portugal to Russia and from Malta to Norway - plus six neighbouring African and Caucasian countries, such as Morocco and Georgia.

Over 60 percent of McDonald's restaurants in Europe are owned and operated by franchisees - local, independent business people. This franchising concept is perhaps the ultimate expression of McDonald's principle of partnership. Following Ray Kroc's idea of "sharing success", it is our business philosophy to develop mutually beneficial relationships with our employees, franchisees, suppliers and the wider community of stakeholders.

The managing directors of each of our national companies are responsible for business operations and results. They report to the Leadership Team of McDonald's Europe Ltd., which is the subsidiary of McDonald's Corporation and is responsible for pan-European business strategies and the guidance and support of country markets. The national companies and franchisees follow common rules on restaurant appearance, menus, supply chain management and operational procedures. The leadership teams in these countries interpret and adapt McDonald's worldwide and pan-European business approaches to reflect their local culture, environment and traditions. This decentralised management approach - encompassing local country management, franchisees and suppliers - creates freedom to innovate within a common strategic framework, learning from each other and sharing best practices internationally, between restaurants, markets and countries.

Creating success in central and eastern Europe: Rocket science not needed!

F.H.J. Schrader
Farm Frites International B.V., Molendijk 108, 3227 CD Oudenhoorn, The Netherlands

Abstract

As for western European (WE) fast moving consumer good producers and distributors (wholesale/retail), it is difficult to increase their sales in their mature domestic markets. Growth strategies have been developed for merging markets like central and eastern Europe (CEE). In CEE the retail environment is changing rapidly to WE standards, while consumer behaviour shows not too many differences. Rocket science is not needed to get on the riding train, but do it now as the train is riding faster and faster...

Keywords: marketing, retail, consumer behaviour, central and eastern Europe

Introduction

This paper will give a draft overview of the retail developments in central and eastern Europe (CEE), some insight into the population and consumer behaviour trends and the key considerations for suppliers to achieve a favourable market position based on the experiences of Farm Frites International B.V.

Central and eastern Europe

With 19 countries, the CEE region - including the Asian part of Russia - covers an area of almost 19 million km2 and has a population of 307 million habitants (CIA, 2005). This overview will focus on the trends of central Europe (Poland, Czech Republic, Slovakia and Hungary) and Russia primarily (Table 1).

The economic growth in CEE remains stable and is above the European Average (2003-2005) of 1.7%, while the inflation is under control - though still high in Russia. The average monthly income is low, though growing considerably with an expected total rise between 30 and 50% in the coming 4 years. However there are considerable social and regional inequalities; *e.g.* in southern Russia the average monthly income is 165 and in Moscow almost 700 Euro. Also, the consumer confidence is considerably higher in the urban areas, where over 65% of the total population lives.

Retail environment and trends

As in western Europe (WE), retailers continue to battle with low single digit growth numbers. International retailers accelerate their investments in Emerging Markets to vitalize their overall growth. In CEE, approximately 1.300 Modern Formats (Hypermarkets/Supermarkets/Discounters)

Table 1. CEE Facts (Source: CIA World Factbook/EIY - Country Forecast/Russian State Statistics Committee). Assumption 35RUR = 1 Euro.

Year 2005	Poland	Czech Rep.	Slovakia	Hungary	Russia
Population in million	38.6	10.2	5.4	10.1	143.4
Area in square km	312,618	78,866	48,845	93,030	17,075,200
GDP in billion USD	489	149	85	159	1,535
GDP/Capita USD	12,700	18,100	15,700	15,900	10,700
GDP Growth	3.3%	4.6%	5.1%	3.7%	5.9%
Av. Monthly Income/Capita*	503 €	694 €	467 €	724 €	183 €
Inflation	2.1%	2.0%	2.8%	3.7%	12.9%

* Figures 2004

are opened each year. These outlets are mainly opened by western European retailers, while in especially Russia some local retailers are growing significantly. Within CEE, hypermarkets have an increasingly dominant position, while hard discount formulas like Lidl and Aldi are now moving eastwards.

In 2004, 9,000 Modern Outlets - shops over 400m2 - counted already for 34% of the grocery sales in the hole CEE region, while over 350,000 traditional stores (and this number is shrinking) were handling the other 66% (Figure 1). Especially in fresh produce the new formats still have to win the hearts of their clients as specialist stores are still winning on price and quality perception; *e.g.* in Poland 53% of the shoppers buy most of their vegetables in specialized shops, while in the Czech Republic this has already been reduced to 12%.

Compared to western Europe, retailers in CEE are less concentrated (Table 2). In Russia, the top 7 retailers - the top 5 are of Russian origin - only cover 9% of the market. We see a tendency

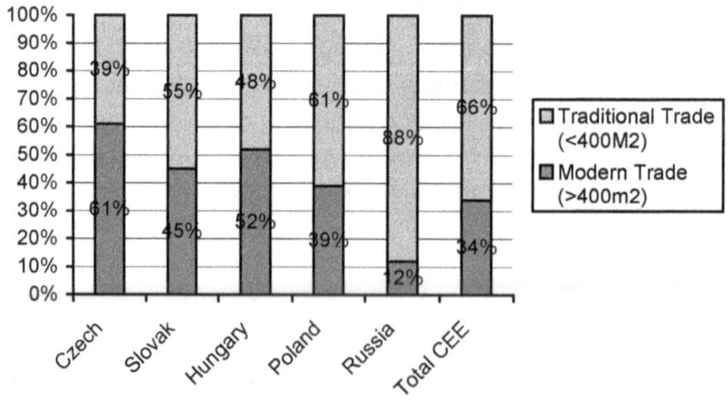

Figure 1. Value share 2004 of packaged grocery sales (ACNielsen).

Table 2. Top 3 retailers and their cumulative market share (ACNielsen).

Country	Number 1	Number 2	Number 3	MS
Sweden	ICA	AXFOOD	KF	91%
Norway	NORGESGRUPPEN	COOP	HAKON	83%
Finland	KESKO	SOK	TRADEKA	80%
Switzerland	MIGROS	COOP	DENNER	77%
Austria	BML-REWE	SPAR	HOFER-ALDI	68%
Belgium	CARREFOUR	DELHAIZE	COLRUYT	68%
France	CARREFOUR	LECLERC/SYST U	INTERMARCHE	64%
Netherlands	AHOLD	LAURUS	TSM	59%
United Kingdom	TESCO	SAINSBURY	ASDA	58%
Ireland	TESCO	DUNNES	SUPER VALUE	58%
Germany	EDEKA	REWE	ALDI	55%
Spain	CARREFOUR	MERCADONA	EROSKI GROUP	54%
Portugal	SONAE	JMR	INTERMARCHE	47%
Slovakia	COOP Jednota	TESCO	REWE	46%
Hungary	TESCO	COOP	CBA	34%
Czech Rep.	MAKRO	AHOLD	REWE	29%
Poland	BIEDRONKA/JMR	TESCO	REAL	13%

that after the market entrance of international retailers a process of consolidation is starting up, especially in those countries where the number of outlets per million habitants is high.

In the whole of Europe, convenience is the main differentiating factor on which basis customers make their store choice. If formulas do not compete on service and/or assortment, price becomes the dominant factor to position a formula which is what the hard discounters make use of.

Consumer behaviour

We can distinguish 3 key consumer trends; health, convenience and value for money. Globally, 75% of the 24 fastest growing categories are related to health. The growth in categories like ready meals and salads combine health with convenience, while private labels benefit from their value for money propositions. Their share is increasing due to a wider presentation in more and more categories and the increased share of international retailers in emerging markets which introduce their international formats and assortment policies. Increasingly, only 1 or 2 A-brands and a cheaper private label of the formula dominate the shelf.

In Russia and to a lesser extend the other CEE countries, the perception of the personal financial status over the next 12 months is positive. 67% of the Russians compared to an average of 57% in Europe, expect an excellent or good personal financial situation. Only 27% of the population will save some cash after covering essential living expenses compared to a global average of 40%. So the consumer sentiment is positive.

The biggest concern of Russians is their health (24%), the development of the economy (24%) and job security (16%), while crime (8%), terrorism (7%) and the political stability (7%) is of lesser concern. Nutrition information on food packaging gets as a consequence more attention than the worldwide average.

New product introductions in western and eastern Europe show that eastern European customers experience the same articles as more unique. However, they are not more positive in purchase intend or liking. Remarkable is that 97% of the grocery shoppers in Russia are female and they love shopping (Figure 2)!

Private label development

The position of private labels will equal western European standards (Figure 3 and 4), though at the moment a huge percentage of consumers in CEE do not entirely trust their quality. International brands are currently more trusted and preferred for their prestige while in some segments local brands obtained historically strong positions.

There are 3 main reasons for the high private label growth in CEE; the increased market share of international retailers following global strategies, discounters playing a more significant role and the increased confidence consumers get in the private label offer over time.

Private labels can gain high market shares in new categories where branded positions are not developed yet (*e.g.* refrigerated ready meals have 47% share in value). Also, food products perceived as more generic (*e.g.* frozen vegetables have 38% share in value) have higher shares as categories with the lowest price distance - brand versus private label - have larger than average private label shares.

In frozen processed potato products, Farm Frites estimates the volume of Private Labels - including distributors brands - already on a level of 40% in the whole of CEE (Figure 5). In value, due to the lower price levels, this is less. Please note that in CEE the price distance between private labels and branded products is often over 40%!

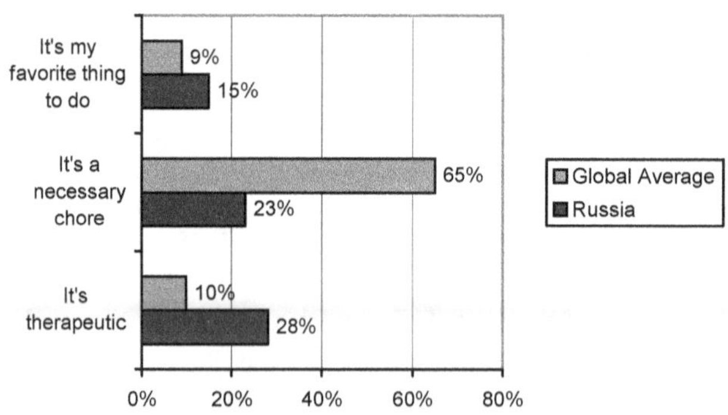

Figure 2. Attitude towards grocery shopping (ACNielsen).

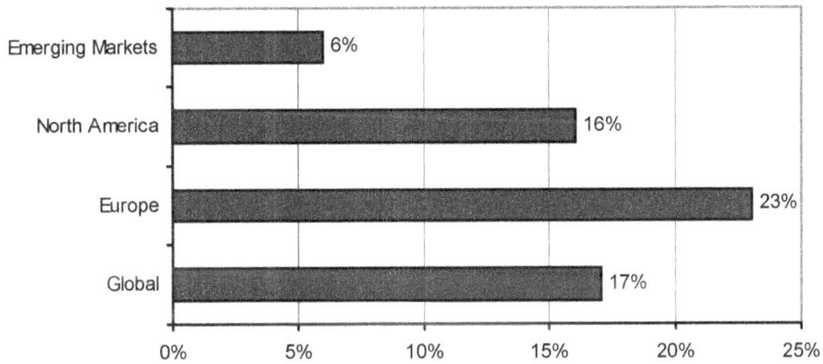

Figure 3. Private Label share in value (ACNielsen).

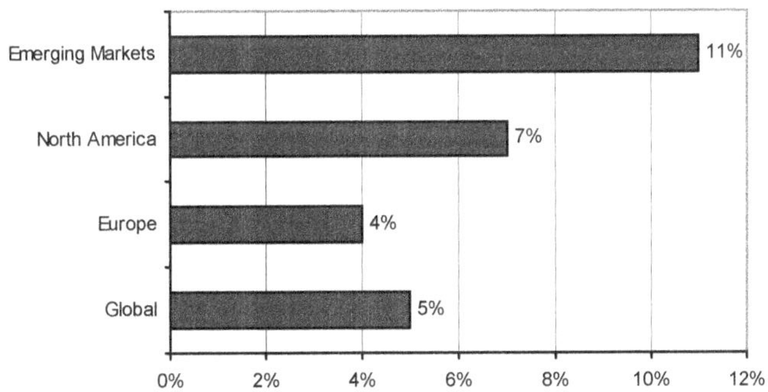

Figure 4. Private Label growth rate in value (ACNielsen).

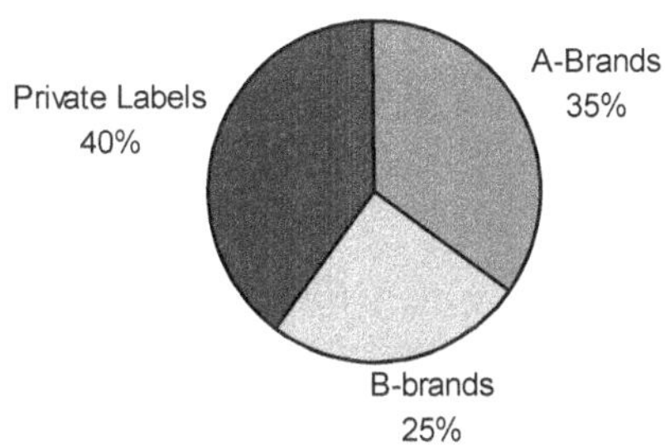

Figure 5. CEE retail frozen processed potato products division in volume (Farm Frites International B.V.).

Farm Frites' position and success factors in central and eastern Europe

Farm Frites was, in 1994 one of the major foreign investors in Poland when it opened it's processing unit and commercial office. Since that time, sales offices were established in Russia (covering also Ukraine and Byelorussia), the Czech and Slovak Republics and Hungary, while an export sales team covers the unmentioned countries. Farm Frites' market share rose to 27% - with an average market growth of 12% per year - in 2005 (Figure 6). Farm Frites expects to sell, in 2006 a volume of 70,000 ton fries and potato specialities - mainly under the Farm Frites brand - in CEE.

The average per capita consumption of frozen processed potato products (excluding international fast-food) is still low with 0.65 kg per year in the whole CEE (Figure 7). While the growth is stagnating in the Czech Republic with a consumption level of 3.5 kg, the Russians are consuming 20% more each year as they consume only 0.2 kg per person per annum. Remarkable is that the segment split between WE and CEE is quite comparable (Figure 8).

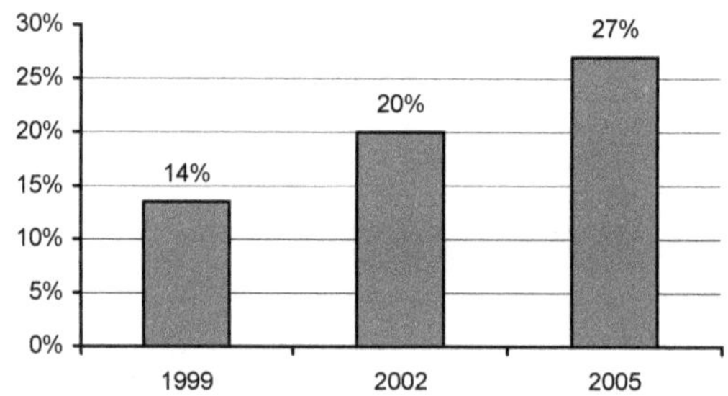

Figure 6. Market share development of Farm Frites in central and eastern Europe (excluding international fast-food) (Farm Frites International B.V.).

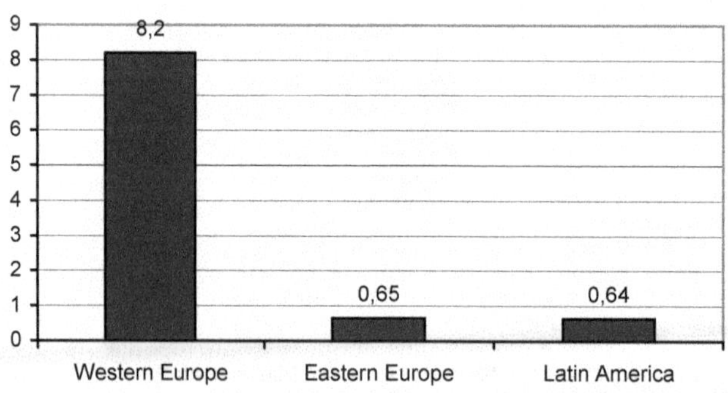

Figure 7. Per capita consumption of frozen processed potato products in kg 2004 (excluding international fast-food) (Farm Frites International B.V.).

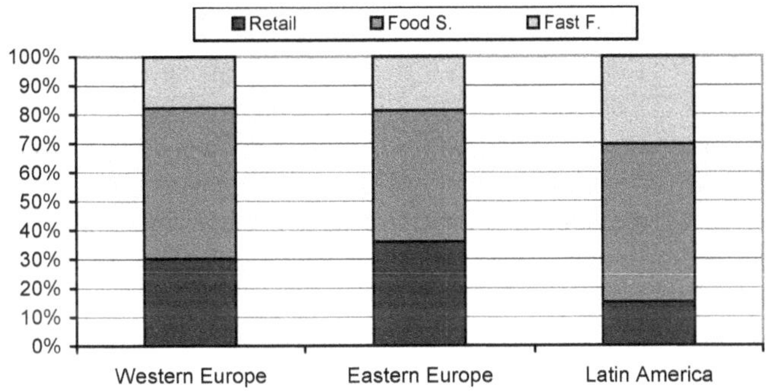

Figure 8. Segment split comparison in volume 2004 (Farm Frites International B.V.).

The following factors were key to Farm Frites' success;
- The creation of a long-term vision (ambition) as a guideline for Farm Frites' strategy. Short term losses are accepted as long as those contribute in building the desired future position.
- Farm Frites strives for long-term partnerships with it's distributors network. We are dedicated to supporting them even if we loose sales on short-term base.
- Farm Frites invests in brand support and it's organisation to achieve the desired market position based on added value propositions.

Conclusions and recommendations

With 307 million habitants and an above average GDP and income growth, CEE is a promising, emerging market but the area is complex. It is enormous in size and the stage of development between the individual countries and urban versus rural areas is huge. As a consequence, building a state-of-the art distribution infrastructure might be difficult.

In retail, we can generally speak of a revolution instead of an evolution in the changes of the distribution structure. International chains are getting a strong foothold in central Europe and even the process of market consolidation takes place. Especially, with the entrance of the (hard) discounters the retail environment equals more and more that of western Europe. Traditional outlets are loosing market share but for some segments continue to be of major importance (especially fresh produce).

There is a positive consumer sentiment and - in many categories - branded positions are not settled yet. There is space for suppliers having an added value proposition and/or can build consumer preference. Key is to build a good distribution position/infrastructure first, which can be complex. High investments should be made in order to become the first or second brand in it's segment as there is no place for a third. While segments can be small, strong positions can be built already and cost inflation is high. The increasing importance of private labels give opportunities for low cost producers with strong value propositions, but limit the growth of the A-brand segments.

We can conclude that eastern Europe offers opportunities belonging to an emerging market, but there are more similarities than differences if we look at the market characteristics and category and consumer trends in CEE compared to WE. Therefore we conclude that Rocket Science is not needed to create success in central and eastern Europe, but simply do your marketing homework and do it better and/or different than your competitors. Segment the market, target the right segments and work on a favourable proposition which CEE consumers want to pay more for or otherwise outperform long-term on costs. But please hurry up as the CEE train is moving forward and is getting too fast to be able to jump on board.

Acknowledgements

VNU/ACNielsen allowed the author of this paper the use of their market data. For more information about ACNielsen and their services we refer to www.acnielsen.com.

References

ACNielsen, 2005: Emerging markets retail and shopper trends 2005

CIA, 2005: http:/www.cai.worldfactbook

VNU, 2006: VNU global spotlight on Russia and Eastern Europe

Curriculum vitae – Frank H.J. Schrader

Frank H.J. Schrader was born in 1970 in Hilversum, The Netherlands. He studied Commercial Economics at the Hogeschool voor Economie en Management (HEAO) where he received a Bachelor Degree in 1991. He holds the post-graduate qualification of NIMA-C. This is the equivalent of a MBA with the Dutch Institute of Marketing (NIMA) and he has joined the AMP Programme of Ashridge, UK.

Since his graduation he worked primarily for companies that produce or market fast moving consumer goods (fmcg), more specifically 'food'. Within Coffeeroaster and Teapacker "Drie Mollen sinds 1818 B.V." and Fruit and Vegetables Cooperative "The Greenery International B.V." he was responsible for various marketing and sales disciplines in both the foodservice and the retail sector.

In 1999, he joined Farm Frites. This is one of world's leading potato processing companies. His responsibility was to grow their branded position in central and eastern Europe (CEE). At the end of 2005, Farm Frites' market share had increased considerably to approximately 30%. In the position of Commercial Director, Mr Schrader is a member of the Managing Board of the Division International Markets (DIM). This business unit covers the full range of company activities (operational and commercial) outside of western Europe. He leads the commercial team of CEE (approximately 100 employees) and his focus is on further developing Farm Frites' position within the CEE and Latin American markets.

Company profile – Farm Frites

Farm Frites, established in 1971, has grown to be one of the world's biggest potato processing companies. The Dutch family owned company - founded by potato farmer Gerrit de Bruijne - processes 1.2 million ton potatoes annually into more than 600,000 ton of frozen and chilled French Fries and Specialities. Farm Frites employs over 1,600 people worldwide, has 8 plants, 16 sales offices and exports it's goods to approximately 45 countries. It supplies to almost all known European retail organisations, international fast-food chains and numerous restaurants close to you.

The organisation of Farm Frites is split into two separate divisions; western Europe and Division International Markets (DIM). DIM sources it's products in five factories in western Europe and makes use of its plants in Poland, Egypt and Argentina, which are located right in the middle of the service areas.

In recent years Sales Offices were established in Poland, Russia, the Czech and Slovak Republics, Hungary and Brazil. While an export sales team covers other areas primarily in central and eastern Europe and Latin America.

Farm Frites is an ambitious grower, which thinks globally. That is why they have a strategic alliance with Simplot, one of the largest food producing companies in the US. Together they are in the top three of the world. Both companies share their knowledge in the field of research and development and the growing of potatoes. Farm Frites believes in doing business personally and wants to be right in the centre of the market. Their strong concentration on creating added value in their products is more and more the pillar of their success.

Issues of the potato industry per country

Potato sector in Poland: From breeding to production

Ewa Zimnoch - Guzowska[1] and Jacek Chotkowski[2]
[1]Plant Breeding and Acclimatization Institute, (IHAR), Młochów Research Center, Platanowa 19, 05-832 Młochów, Poland
[2]Plant Breeding and Acclimatization Institute, (IHAR), Bonin Research Center, 76-009 Bonin near Koszalin, Poland

Abstract

Potatoes belong to the main crops in Polish agriculture, as potato consumption in this country reaches 129 kg per capita. Breeding of cultivars is realised in three breeding companies. In the Polish National List 132 potato cultivars are registered, 80 of which are domestic, and 52 are of foreign origin. Polish cultivars dominate among registered starch cultivars. That group possess also a high level of resistance to pathogens (viruses and late blight). Cultivars of Dutch and German origin are predominantly used for crisps and French fries production. Table cultivars of domestic origin are slightly weaker in quality traits than comparable group of foreign cultivars of. A significant decline of qualified seed production was noted in Poland during last 15 years. In 2005, qualified seeds were produced on 4 733 ha. Since 1990 the potato production area has declined by 1.2 million ha. In 2005 potatoes were grown on 588 thousand hectares with 10.4 million tons harvested, of which 36.5% were marketable. The average potato yields are low, what results from several reasons, of which high diversification of potato production among small farms with traditional farming system seems the most important. It is followed by traditional use of a significant part of potato yield (34.1%) for fodder. Processing of potatoes has an increasing importance in recent years, however only 12.8% of total yield is used for it. From 1.8 million tons of processed potatoes in 2004/2005 season about 800 thousand tons were processed for crisps and French fries and 760 thousand tons for starch. A low input type of technology predominates in Polish potato production. The average input of fertilizers is about 250 NPK kg/ha and the average use of chemicals is 3.5 kg/ha of active compounds. The Polish potato sector adjusted its regulations to the EU needs.

Keywords: Poland, country profile, cultivars, production, crop management, processing, trade

Breeding

Besides cereals and sugar beet potatoes, *Solanum tuberosum* L. belong to the main strategic crops in Polish agriculture. Commercial breeding programs are realized in Poland by three breeding companies. Two companies are owned by the Agricultural Property Agency: The Pomeranian and Mazovian Potato Horticulture Company in Strzekęcin in which more than 40 cultivars were bred, and the Plant Breeding Company in Szyldak with 11 cultivars. The third one Zamarte Potato Breeding - Group IHAR is owned by the Plant Breeding and Acclimatization Institute. This company was created in 1946 and more then 70 cultivars were bred there. This structure of potato breeding companies has resulted from structural changes among nine breeding stations existing in Poland after 1945.

Cultivars

In 2006, a total number of 132 potato cultivars are described in the Polish National List, 80 of which are domestic, and 52 are registered by foreign companies (Dutch and German). Polish cultivars dominate among registered starch cultivars, with mid-late or late maturity. In the groups of table cultivars or those suitable for processing the representation in the register is balanced; however the domestic cultivars are rarely used for crisps and French fries processing. Prevalence of mid-early cultivars is noted in both groups. The structure of registered cultivars according to their maturity type is shown in Figure 1.

On average, cultivars of Polish origin possess a high level of resistance to pathogens, mainly viruses and late blight. It is worth noting that 20% of the domestic cultivars are extremely resistant to PVY, the most devastating viral pathogen, which has recently, besides a yield decrease, evoked strains causing necroses on tubers. Among the registered Polish cultivars there is a good representation (about 20%) of cultivars expressing resistance to late blight. All cultivars registered in Poland are wart resistant. This is a 10-year obligation for cultivars registered in Poland stemming from Poland's Membership in the EU. An exception is made for wart susceptible cultivars which are produced for processing purposes under special permission of the Main Inspector of the Plant Health and Seed Inspection Service confirmed by the Council (for one-year cultivation, on the fields known as free from the pathogen). Such permissions were given 77 times in 2005 for 4600 ha of potato production (*e.g.* for cultivars Hermes, and Lady Rosetta).

Table cultivars of domestic origin are on average slightly weaker in quality traits than comparable group of cultivars of foreign origin. Data of selected quality traits are shown in Table 1. Nonetheless, when resistance to main pathogens is taken into account, the higher resistance level of domestic cultivars is evident. It might be concluded to be a result of the agro-climatic conditions in which breeding and cultivation take place in Poland. The long-term activity of parental lines breeding program held by IHAR, from which more than 50 Polish cultivars were selected and grown in the country might be seen as a result. In Figure 2 the mean resistance to PVY, PLRV and late blight is presented for cultivars registered in 2002 in Poland and grouped into three classes in relation to their origin.

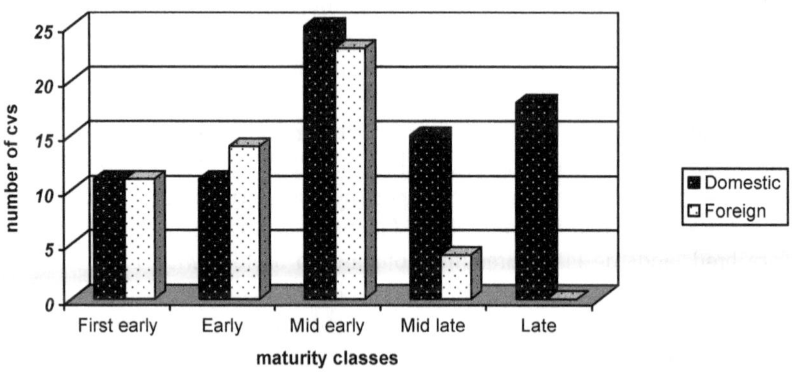

Figure 1. Maturity classes of potato cultivars registered in the Polish National List in 2006.

Table 1. Means and range for the evaluation of quality traits of table cultivars registered in 2005 in the Polish National List (Chotkowski and Stypa, 2005).

Cultivars	Number	Means and range of evaluated traits[a]		
		Taste	Enzymic browning	After cooking blackening
Foreign origin	43	7.0	7.6	8.3
		6.5 - 8.0	7.0 - 8.5	7.5 - 9.0
Domestic origin	49	6.8	7.8	8.2
		6.0 - 8.0	7.0 - 8.5	7.5 - 8.5

[a]Evaluation in 1-9 grade scale, where 9 = the best

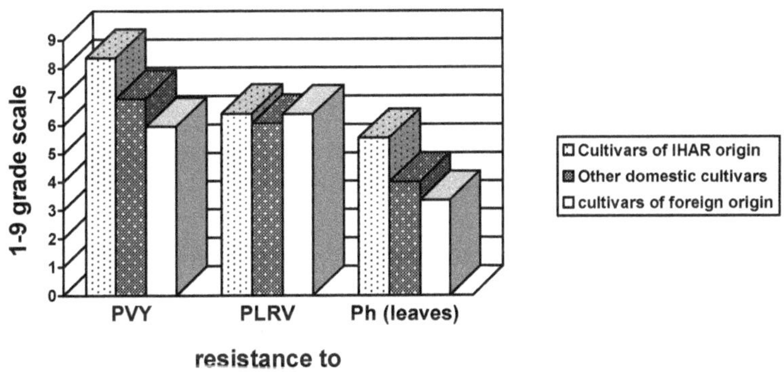

Figure 2. The mean resistance to PLRV, PVY and late blight of cultivars registered in Poland grouped according to their origin: domestic cultivars originated from IHAR's parental lines, other domestic cultivars and cultivars of foreign origin.

Seed production

A significant decline of qualified seed production was noted in Poland during the last 15 years. It resulted from a decrease in potato cultivation area, very low seed exchange rate (about 5%), which is a historical problem related to a large percentage of not market-oriented small potato producers among Polish farmers, as well as weakness of the potato markets. Throughout the last two seasons the development of the potato crop was inhibited because of an occurrence of bacterial ring rot in the country.

In 2005, qualified seed was produced for 127 cultivars (including 16 cultivars not registered in Poland) on 4,733 ha. The largest seed production acreage was for the cultivar Vineta (364 ha). Table 2 demonstrates acreage and indices of seed production in Poland since 1990 (data from Plant Health and Seed Inspection Service).

Table 2. Seed potato production in Poland between 1990 and 2005 (www.piorin.gov.pl).

Year	Acreage of seed production (ha)	Input of seed acreage in total potato production (%)	Input of qualified seeds in planted seeds (%)
1990	46,540	2.53	5.6
1991	19,909	1.13	4.7
1992	16,281	0.93	3.5
1993	13,602	0.77	4
1994	10,300	0.61	2.7
1995	8,117	0.53	3.7
1996	12,404	0.9	2.4
1997	10,640	0.81	4.2
1998	9,524	0.74	3.3
1999	7,446	0.59	3.4
2000	7,012	0.56	2.5
2001	6,975	0.58	2.6
2002	7,097	0.88	3.9
2003	6,399	0.84	4.2
2004	6,071	0.85	5.1
2005	4,736	0.8	5.4

Since Poland has joined the EU, the potato production rules were changed especially for seed production. Several new rules have been created for registration of producers, potato passports, quarantine regulations, certification forms, *etc.* It reflects also in changing laws for cultivars protection and farmers' saved seed.

Crop production

In 2005, despite decreasing tendency, potatoes were cultivated on 588 thousand hectares that was equal to 5.3% of the total sown area of crops. Since 1990 the potato area has declined by 1.2 million ha. Potatoes are grown in all Polish regions. Gross potato yield in 2005 was at the level of 10.4 million tons, 3.8 million tons of which were sold on the markets (Table 3). Input of market oriented potato production was estimated on the level of 36.5% of gross potato yield, while until 2002 it was lower than 20% of total yield.

During the last 25 years the average potato yield in Poland was about 17.4 t/ha, that was equal to about 50% of corresponding potato yields in Germany or in the Netherlands. Low yields in Poland resulted from several reasons: a very high number of small potato fields, and a not well developed system of crop management: lack of irrigation of potato fields and improper crop protection for the majority of potato fields, and frequent use of non qualified seeds by smaller farmers. More than 50% of potato fields in Poland are below 1 ha in size. In 2002, the average size of a potato field was 0.52 ha for 1.5 million of Polish farms, where potatoes were grown. Only 6,900 farms in Poland belong to the groups having potato fields over 5 ha in size. In these groups of farms potato is harvested from 74,000 ha (Table 4).

Table 3. Potato production in Poland 1990-2005 (Rocznik Statystyczny 1991, 1994, Statistical Yearbooks of the Republic of Poland 1998, 2004, www.stat.gov.pl *).*

Year	Area harvested	% of total sown area	Yield	Gross yield	Marketable production	
	1,000 hectares	%	t/ha	million tons	million tons	% of gross yield
1990	1835	12.9	19.8	36.3	6.4	17.6
1991	1733	12.2	16.8	29	5	17.2
1992	1757	13	13.3	23.4	4.2	17.9
1993	1761	13.1	20.6	36.3	5.2	14.3
1994	1697	13.1	13.6	23.1	3.7	16
1995	1522	11.8	16.4	25	3.8	15.2
1996	1342	10.9	20.3	27.2	4.8	17.6
1997	1306	10.5	15.9	20.8	3.7	17.8
1998	1295	10.3	20	25.9	4	15.4
1999	1268	10.1	15.7	19.9	3.2	16.1
2000	1251	10.1	19.4	24.2	4	16.5
2001	1194	9.6	16.2	19.4	3.6	18.6
2002	803	7.5	19.3	15.5	4	25.8
2003	766	7	17.9	13.7	3.9	28.5
2004	713	6.3	19.6	14.0	3.8	27.1
2005	588	5.3	17.6	10.4	3.8	36.5

Table 4. Acreage structure of potato production according to the size of farms growing potatoes in Poland, 2002 (Central Statistical Office, GUS , data of the Agricultural Census 2002).

Groups of farms acc. to the size of potato fields ha	Potato fields			Number of farms	
	Total, ha	% of potato acreage	Per farm, ha	Total	% of all farms
< 1	406,833	50.6	0.3	1,328,544	85.4
1 - 2	196,323	24.5	1.2	169,430	10.9
2 - 5	128,916	16	2.6	50,302	3.2
5 - 10	33,359	4.2	6.2	5,423	0.3
10 - 15	9,844	1.2	11.2	882	0.1
15 - 20	3,569	0.4	16.1	222	0
20 - 50	7,269	0.9	30	280	0
> 50	17,284	2.2	147.7	117	0
Total	803,385	100	0.52	1,555,200	100

Usage of potato yield

The estimated structure of the usage of the 2004 potato yield is shown in Table 5. The estimated demand for food potatoes in 2004/2005 season is about 4.4 million tons equal to 32% of total yield. Almost the same demand is for fodder. Data of the last decade indicate a decrease in the use of potatoes for fodder, as in 1997/1998 about 42% of 20.8 million tons produced potatoes were used for feed.

In the 2004/05 season the total amount of potatoes processed by the industry irrespectively of the purpose was about 1.8 million tons. In recent years lower amount of processed potato were noted in 1997 and 1999, respectively when lower yields were obtained. The industrial processing for starch in 2004/2005 season was lower by 219 thousand tons than in the previous campaign. It reflected the ceiling of starch production, which was restricted to a national quota of 145 thousand tons. Poland is very interested in the increase of the starch quota by the EU. The processing for ethanol and dried products did not indicate significant changes during the last two seasons. The interest of the EU in the enlargement of ethanol production for bio-additives to the gasoline might change this type of potato use in the near future in several EU countries including Poland. An increasing tendency is clearly noted for potatoes processed into crisps and French fries. From 246 thousand tons in 1997/98 the amount of potatoes for processing increased to 645 thousand tons in 2004/05 (Table 6).

Poland is one of the leading countries in potato consumption. In 2004/2005, the per capita consumption was about 129 kg potatoes. This amount has slightly decreased in recent years, as in 1997/1998 the consumption was on the level of 134 kg per person; however, the increasing input of the consumption of chips and crisps is noted from year to year: from 8.9 kg of potatoes used per person for processing in 1997/1998 to 13.1 kg used in 2004 (Table 7).

Table 5. Estimated structure of the use of potato yield harvested in 2004 (data of Central Statistical Office, Dzwonkowski et al.*, 2006).*

Type of use	Thousand tons	%
Food	4,420	31.4
Fodder	4,791	34.1
Processing	1,792	12.8
Seed	1,480	10.5
Export	30	0.2
Losses	1,550	11.0
Total	14,063	100

Table 6. Potato processed in Poland between 1997-2004 in thousand tons (Dzwonkowski et al., *2003, 2006).*

	1997/98	1998/99	1999/00	2000/01	2001/02	2002/03	2003/04	2004/05
Starch	470	806	355	815	690	910	979	760
Ethanol	212	223	83	215	100	100	86	102
Dried products	212	233	140	175	85	85	133	130
Crisps, French fries, others	246	317	350	450	530	600	543	800
Total	1,140	1,562	908	1,655	1,405	1,695	1,741	1,792

Table 7. Potato consumption per capita in Poland (in kg/person) (Dzwonkowski et al., *2006).*

	1997/98	199899	1999/00	2000/01	2001/02	2002/03	2003/04	2004/05
Total	134	135	131	132	130	131	130	129
As processed potatoes	8.9	10.7	11.4	11.4	11.9	12.6	12.8	13.1

Potato production technology

The technical equipment applied for potato production varies in relation to use the production sector. Głuska and Nowacki (2005) discriminated between three production targets: on-farm use, fresh market and starch, and processing. In Table 8, differences are shown in the application of fertilizers, weed control, use of irrigation, stone removal, row spacing and predominant storage method.

A low input type of technology predominates in Polish potato production, resulting from the significant input of small, not market-oriented farms in the structure of all farms growing potatoes. The average input of fertilizers is about 250 NPK kg/ha and the average use of chemicals is 3.5 kg/ha of active compounds. Potato production for processing into crisps and French fries is located in modern large-sized farms. Applied technology and planted cultivars in such farms are based on signed contracts to the processors. This group of producers use a high input technology including irrigation of potato crop. Large doses of fertilizers are combined with chemical crop protection by 10-12 sprays per season as a guarantee of high quality yields.

Foreign trade

In the 2004/2005 season the balance of foreign trade of potato and potato products has declined. After the accession of Poland to the EU the imports of potato products (mainly starch) have increased at a relatively high rate. The same was noted for exports of potato products,

Table 8. Elements of potato technology in different production sectors (Głuska and Nowacki, 2005).

	Production targets		
	On-farm use	**Market and starch**	**Processing**
Fertilizers			
- NPK (kg/ha)	50-100	180-400	550-650
- manure (% of farms)	95	65	occasionally
Weed control (% of farms)			
- mechanical	100	22	3
- chemical	0	10	85
- combined	0	68	12
Plant protection treatments (#)			
- fungicides	0-1	1-5	8-12
- insecticides	0-1	1-2	1-3
- sprout suppressants	0	5	50
Irrigation (% of farms)	0	5	35
Stone removal (% of farms)	0	Occasionally	25
Row spacing (% of farms):			
- 62.5-67 cm	100	90	10
- 75-90 cm	0	10	90
Basic storage method	Clamps, cellars	Clamp, adapted buildings	Professional storages

however at a lower rate. Ware potato export declined in 2004/2005 and 2005/2006 seasons, comparing to the 2002/2003 and 2003/2004 exports (Table 9).

The Polish potato sector adjusted its regulations to the EU needs. The forecast for potato production in Poland in the coming seasons is not expecting major changes. However, in a medium run the potato acreage will tend to decrease to the level of 500 thousand ha and seed acreage will increase. Polish potato cultivars might play a significant role for the European starch industry and for ecological production.

Table 9. Foreign trade of potato and potato products in 1999- 2005 in thousand tons (Dzwonkowski et al.*, 2003, Dzwonkowski* et al.*, 2006).*

	1999/00	2000/01	2001/02	2002/03	2003/04	2004/05	2005/06[a]
Exports:							
Fresh potato	57.9	83.6	19	139.9	108.8	30.4	20.9
Potato products	93	142.8	151	169.9	204.8	224.0	265.2
Total, equivalent to fresh potato	476	697	696	924	1064	1097	1264
Total exports in million €	58.2	80.9	86.6	93.1	108.6	122.7	152.0
Imports:							
Fresh potato	131.1	39.5	70.9	29.8	35.9	90.3	84.0
Potato products	75.7	88	88.7	98.5	114.3	225.1	264.0
Total, equivalent to fresh potato	575	559	594	612	709	1386	1637
Total import in million €	60.2	52.6	54.3	56.1	57.8	120.3	134.0
Trade balance in million €	-2.0	28.3	32.3	37.0	50.8	2.5	18.0

[a]forecast

References

Anonymous. Central Statistical Office. www.stat.gov.pl

Anonymous. Plant Health and Seed Inspection Service. www.piorin.gov.pl

Głuska A. and W. Nowacki (2005). Technology of potato production in Poland: Workshop "Potato production and storage for good tuber quality", Warsaw, March 01-04, 2005 Centre of Excellence CICSA, IHAR: 71-77

Chotkowski J. and Stypa I. 2005. Odmiany ziemniaków. Charakterystyka tabelaryczna. www.ihar.edu.pl

Dzwonkowski W., Szczepaniak I., Rosiak E., Chotkowski J., Rembeza J. (2003). Rynek Ziemniaka stan i perspektywy. Analizy rynkowe. IERiGZ. No 23

Dzwonkowski W., Szczepaniak I., Rosiak E., Chotkowski J., Rembeza J., Bochi ska E. (2003). Rynek Ziemniaka stan i perspektywy. Analizy rynkowe. IERiGZ. No 24

Dzwonkowski W., Szczepaniak I., Rosiak E., Chotkowski J., Rembeza J., Bochi ska E. (2004). Rynek Ziemniaka stan i perspektywy. Analizy rynkowe. IERiGZ. No 25 and 26

Dzwonkowski W., Szczepaniak I., Rosiak E., Chotkowski J., Rembeza J., Bochi ska E. (2006). Rynek Ziemniaka stan i perspektywy. Analizy rynkowe. IERiGZ. No 29.

Rocznik Statystyczny.GUS. Warszawa, 1991, 1995

Statistical Yearbook of the Republic of Poland. GUS. Warsaw. 1998, 2002, 2004

Curriculum vitae – Ewa Zimnoch-Guzowska

Ewa Zimnoch-Guzowska is a Professor in Agronomy (Plant Breeding) at the Plant Breeding and Acclimatization Institute (IHAR), Poland and Head of the Młochów Research Center of IHAR, where she leads the Department of Potato Genetics and Parental Lines. She received a M.Sc. degree in Horticulture from the Warsaw Agricultural University in 1970, and than she joined the research group of Professor B. Kubicki at the same university to study sex determination of higher plants. In 1979 she received a Ph.D. degree in Agricultural Science. In 1982 she moved to the Potato Research Institute (merged to IHAR in 1983), where she was in charge of Laboratory of Genetics focused on exploration of *Solanum* species in potato breeding. She was a successor of Professor K. wie ynski and since 1992 she is in charge of the Młochów Research Center. She got the D.Sc. degree in 2001 and was nominated to a Professor of Agronomy by the President of Poland in 2005. During 1999-2004 she held the position of a Visiting Adjunct Professor in Department of Plant Pathology, Cornell University, Ithaca, USA.

The main research interest of Professor Zimnoch-Guzowska was associated with the potato parental line breeding at diploid and tetraploid levels. Her particular interests were focused on exploitation of *Solanum* germplasm for breeding lines outstanding in starch content and resistance to *Erwinia*, lines resistant to late blight in foliage and tubers with shorter vegetation period; research on *Phytophthora infestans* resistance sources searching for durable resistance; mapping potato genome - mainly genes for resistance to pathogens and the use of molecular markers in potato selection process.

She was awarded three times a collective price for research from the Minister of Science and High Education in Poland in 1977 and 1980, and in 1993 from the Minister of Agriculture and Food Economy. She is a member of national and international scientific bodies and participates in various networks and projects that deal with the breeding research on potatoes. Since 2002 she acts as a chair of the Potato Section of the Polish Seed Association.

Company profile – Plant Breeding and Acclimatization Institute (IHAR)

IHAR was founded in 1951 as a scientific unit focused on research in breeding and seed production of major field crops. IHAR is the largest agricultural research institute in Poland. The Institute's headquarters are located in Radzikow. Research centres and experimental stations of IHAR are located in several regions of Poland. Centres involved in potato research are located in Bonin, Jadwisin and Mlochów.

In the IHAR Mlochów Centre research is focused on potato genetics and breeding research. Genetic studies are determining mode of inheritance of selected potato characters, like resistance to viruses, late blight and some quality traits. Potato resistance traits determined by major genes and QTL's are mapped with an international cooperation. Preliminary studies are conducted on marker assisted selection (MAS) and on the application of PCR for diagnostics of pathogens. Novel sources of resistance to viruses and late blight are tested. Studies on the biology of potato pathogens, including viruses, with special attention to their new pathotypes are being carried out. The variation in population of potato viruses (mainly PVY) has been investigated. In Mlochów all cultivars applying for the registration in the Polish National List are evaluated for virus resistance. A collection of unique diploid and tetraploid potato germplasm and collection of potato pathogens (viruses and *P. infestans*) are maintained at the Mlochów Centre. Improvements in diagnostics, germplasm storage and screening methods are important part of germplasm evaluation and technology development studied in Mlochów.

Parental lines are bred at IHAR Mlochów for all Polish potato breeding companies. In their development, special attention is paid to table and processing quality, starch content and resistance to main potato viruses (PVY, PLRV, PVM, PVX, PVS), *Phytophthora infestans*, nematodes and wart. Yearly, since 1967, selected parental lines have been offered to Polish breeders. Over 50 cultivars registered in Poland originate from Mlochów parental lines.

Current trends in the Hungarian potato production

E. Zsom[1] *and J. Kruppa*[2]
[1]*Szíriusz Trade Bt., H-1126 Budapest, Dolgos u. 2. Hungary*
[2]*University of Debrecen, Centre of Agricultural Sciences, Faculty of Agriculture, Department of Crop Production and Applied Ecology, H-4032 Debrecen, Böszörményi út 138. Hungary*

Abstract

The most important trends of the potato industry in Hungary, which have particularly received scientific attention in recent years, are described. Special emphasis was given to the market possibilities, cultivar use and the current level of technology, while also highlighting the major impacts of Hungary's EU accession on the domestic potato industry. Hungary has a less favorable situation for potato production than countries with a more northern, north-western location. Our climate is favorable primarily for growing early and virus-resistant cultivars. Regarding the soil conditions, the loose sandy soils of the three great sand regions (Duna-Tisza közi homokhát, Nyírség, Belső-Somogy) are the most suitable for the production of high-quality, mainly early and summer potato. Due to the free flow of potato produced in the EU, the production of early and virus-resistant cultivars will be competitive and profitable in Hungary, which enables the production of cheaper and better quality seed potato, which could result in as much as a 50% cut in seed potato prices. Consequently, the production and prime costs of table potato will also be reduced. Regarding skin coloration, Hungarian customers have been averse to cultivars with yellow skin, a behaviour which seems to change. Among those yellow-skinned cultivars, Agria, Hópehely and the early Pannónia are the most accepted ones.

Keywords: *Solanum tuberosum*, Hungary, cultivars

Growing area

A reduction in the potato growing area can be observed in Hungary since the last 40 years. This reduction has continued since 1999, and presently this area (23,800 ha) is 50% smaller than in 1999 (Figure 1). Additionally, the area for seed potato growing has reduced to one third of that it was when compared to the two years before, to 430 ha.

Yields and consumption

Although the production area and the willingness to grow potatoes are reducing, yields have increased. Yields have been improved for several decades, and the national average yield is now 25.3 t/ha. This average safely provides the necessary 550-600,000 t annual total yield, which is enough to satisfy the domestic consumer demand. Problems are experienced only if there are great losses due to potato blight or drought, or *e.g.* if the storage conditions are worse than usual. In these cases, table potatoes should be imported to ensure the continuous supply and to enlarge the product assortment. In Hungary, the per capita potato consumption has fallen to about half of that in the 1930s, from 130 kg to 60-65 kg/person/year, and has been stable for the last few years (Figure 2). The reduction in consumption was caused not only by the

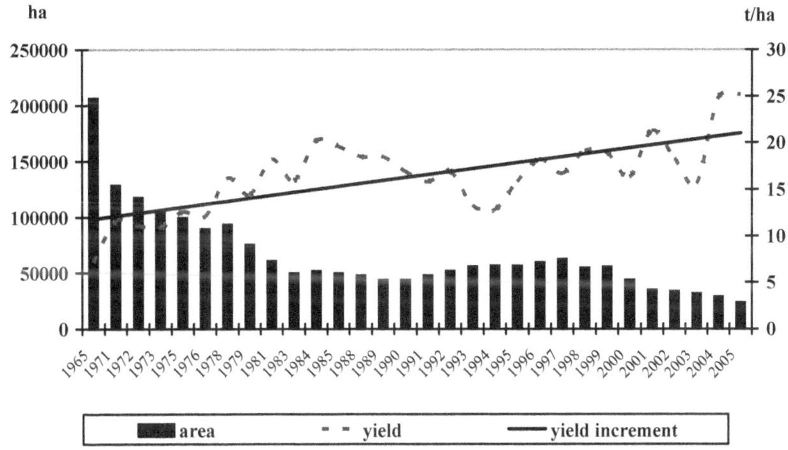

Figure 1. Changes in the growing area and yield of potatoes in Hungary in the period of 1965-2005 (data of the Central Statistical Office (KSH, 2006)).

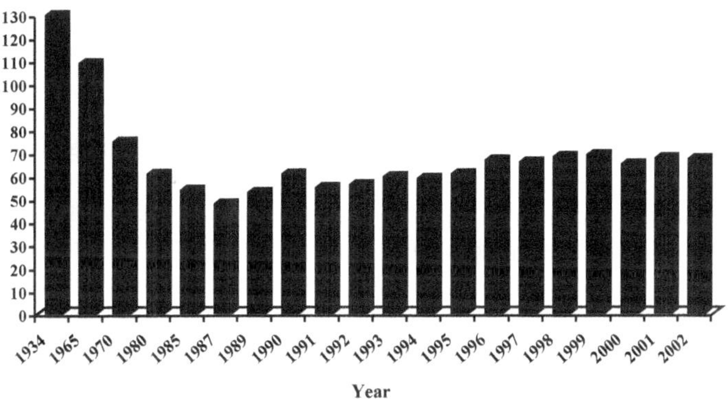

Figure 2. Per capita potato consumption in Hungary (kg) in the period between 1934 and 2002 (based on the data of CSO (KSH, 2003), Food and Agriculture Organisation (FAO, 2005) and the National Potato Product Association (OBTT, 2003)).

improving living standards, but also by the changes in consumer behavior, by high prices in some years and the frequently inferior quality.

The dynamic development of the domestic potato processing industry would be necessary, which would enable the enlargement of domestic production. As an example, the production of deep-frozen fried potatoes stopped some years ago. The ratio of processed products is low: less than 5% of the produced potatoes are sold in a processed form. The main reason for this is the lack of consumers who could afford such foods.

Cultivar assortment

Since the mid-1970s, the cultivar assortment has consisted of foreign cultivars. Seed potato supply has been based primarily on the import of seed potatoes, mainly of Dutch cultivars, and their annual production. Due to the appearance of an aggressive strain of the potato Y virus at that time, the old, virus-susceptible Hungarian cultivars have disappeared from the production. Hungary changed to the production of other cultivars, which are also susceptible to the virus, but can be propagated safely in countries with favorable conditions. These Dutch cultivars also deteriorate in Hungary in one or two years, just as the old Hungarian cultivars did. Therefore, we are dependent upon import seeds each year, which makes the production exposed and expensive. Since there has not been a virus-resistant Hungarian cultivar on the market for almost 25 years, the foreign cultivars (Desirée) and their distributors had no competitors. The attitude that was formed and stabilized in the past decades, *i.e.* the Hungarian consumer's aversion to yellow-skinned cultivars, has reduced the cultivar assortment, and the market has been dominated by only three red-skinned Dutch cultivars (Desirée, Kondor, Cleopatra) for the last 10-15 years. In recent years, several growers were forced by the economic conditions to grow cultivars from Keszthely, which have cheaper seed and can be grown more profitably, in order to remain competitive (Polgár, 2006). As a consequence, the three red Dutch cultivars (Desirée, Kondor, Cleopatra), the Keszthely cultivars (Százszorszép/Rioja/, Góliát, Hópehely), the yellow-skinned German cultivar Agria, and the other newly introduced cultivars (yellow-skinned Pannónia) have a domestic market share of 70%, 18%, 10% and 2%, respectively.

The dry matter content of the Keszthely cultivars is higher than that of the foreign cultivars grown in Hungary. They have a good taste, can be cooked more quickly and the tubers have a better appearance. Since they are more resistant to diseases, mainly to viruses, than foreign cultivars, their seed potato can be produced in Hungary and the profitability of their production is competitive with that of foreign cultivars.

The use of cultivars is determined by the size of the farm, the larger the area the higher the number of cultivars used (they have possibilities to test cultivars). From the frequency of the cultivars, it can be concluded that the Dutch cultivars are absolutely prevalent. The most frequently planted cultivar is Kondor, although the number of farms choosing this cultivar is decreasing. The area of Cleopatra is of the same order, but it is also diminishing. Cultivars with medium planting frequency are Desirée, Kuroda and Agria. In the first places there has not been a change yet, but if the present trend continues, then new cultivars can prevail over Kondor, such as Kuroda and Agria. The other cultivars are scarcely planted. Presently, only a small number of farms use the Hungarian Keszthely cultivars, Százszorszép (Rioja), Góliát and Hópehely have the largest shares.

If we study the data on the different maturity groups, we can see that the smaller farms tend to grow early cultivars, while the larger farms produce mainly later ones. The smaller farms usually grow several cultivars from the different maturity groups, while the larger growers usually grow less cultivars.

Market competition

As a result of the stronger market competition which followed the EU accession, Hungary has lost its markets for autumn and stored table potato, since the countries of northern and northwestern Europe offer good quality table potato in this period at a more favorable price. They are able to do so, because they have significantly lower prime costs due to the more favorable (cooler, more humid) climate for potato production. Accordingly, in order to improve its competitiveness and to keep its markets, Hungary has to reduce the prime costs considerably, since potatoes are produced too expensively, with a 20-25 HUF/kg prime cost, except for some companies producing at a high level. The prime cost of table potato should be reduced to 12-14 HUF/kg (Kruppa, 2005). Early potatoes may have higher costs due to the generally higher market price.

In each year, prices are strongly influenced by the year, the produced amount and the general health of tubers. In years with high disease pressure, like in 2005 with potato blight, the growers try to sell the potatoes in autumn, even at a loss. Consequently, table potato have to be imported in the spring, which has an impact on the domestic prices. Growers have an increasingly active role in influencing the prices. The washed and packed potatoes can be sold at a higher price. In the supermarket chains, selling at a lower price than the prime cost can be seen, which interferes with the growers' interests.

Production technology and seed

Regarding production technology, the basic criterion of competitiveness is the development of irrigation, nutrient supply and storage capacity. The 250,000 t storage capacity of Hungary should be increased by a further 200,000 t, in order to ensure the conditions for the safe storage of the potato produced in Hungary within a short time.

When studying the questions of machinery, it can be stated that most of the small and medium growers have the necessary equipment for soil preparation and other processes. The special machineries used for potato growing are usually missing. Even if they own such machines, they are German and Dutch ones from the 1970s, which were scrapped from the EU, with a performance much lower than those used by western European competitors. On most of the farms, German potato combine-harvesters and Czech planting machines are still used, the average age of which is 20-25 years. However, in addition to its effect on profitability, the technological level also determines the quality of the product.

As regards the use of seed tubers, healthy (of certified origin) seed tubers are used only on 30% of the total growing area. About 60-65% of those using certified seed tubers change the propagation material each year, while the others change it in 2-3 years. The purchase of seed potato from the seed firms is done via individual agreements. Only 5-6% of the table potato is grown from direct import.

Many are dissatisfied with the quality of seed potato, they claim that it is not uniform and is of medium level. Due to its significance and high price, growers purchase the seed tubers only from reliable distributors. This product requires trust, therefore, a stable partnership forms between the seed potato growers and their customers. Buyers of seed potato tend to purchase from breeders directly, since the Hungarian breeders usually sell also smaller amounts and give advice on the technology. Most of the growers consider the price of seed tubers high, but they know that this is due to the low yields.

Market

In the improvement of the distribution system's efficacy, a gradual increase in the production based on contracting should be targeted. This can be achieved by developing the existing relationships with the integrator networks and commercial networks. A basic criterion for this is to satisfy the market and consumer requirements. In order to achieve this goal, technological, commercial, marketing and structural tasks should be completed, in which all the industry's participants should be involved.

In the market, a higher price can be obtained for the potato stored during the winter and sold in the spring. After Hungary's EU accession, higher quality requirements were set for potato due to the strong market competition. We can acquire new markets (market niches) with early yellow-skinned potato mainly in northern countries (*e.g.*, Estonia, Latvia, Lithuania), if the Hungarian potato can arrive in these countries at the appropriate time and can satisfy the quality requirements at a competitive price.

In the western European markets, the early potato already appears at the beginning of December, mainly from Egypt, Tunisia, Morocco and Israel, then from Malta and Cyprus, and finally from Portugal, Spain and Italy, until the domestic potato becomes available. Although the Hungarian early potato has a much better quality and taste than the above-mentioned southern potato, which is also even stored sometimes for several weeks and months, it cannot be competitive with them in the earliest period. However, from May on, the Hungarian consumers prefer the domestic fresh and tasty early potatoes, and we have marketing possibilities in the northern European markets from June, since we can be competitive due to the shorter distance.

Among the market participants, the small growers are the most defenseless and their expertise is poor. Most of these growers produce potato occasionally and do not have market contacts. The amount marketed by small growers is around 120-140 thousand tons. Although this accounts for 20-25% of the total yield and has a serious role in price determination, it appears on the market in a short period around harvest.

Modern, large farms have the necessary machinery, therefore, they can achieve higher yields and have a high profit. They additionally possess enough financial sources to make investments and are able to operate integrated networks.

There have been considerable changes in the servicing function of large-scale farms. Formerly, large-scale farms delivered extensive services (planting, hilling, desiccation, harvesting), but since the appearance of the bacterial disease Ralstonia, large-scale farms have stopped the delivery of such services, and they do not rent their special machines due to the plant health problems. Large-scale farms have an important role in determining market conditions (demand, supply, prices), in the development of the production technology and in the spread of modern methods.

The growers are defenseless against the buyers, since most of the buyers sell the products on the wholesale market. Usually, they do not make a contract with a grower in advance. The grower has to finance the production process and will get the money only after selling the product. In the grower's opinion, there is little chance to fulfill the contract, since the amount of yield is dependent upon the weather and it is either the grower or the buyer whose interest is opposite to the contract. In those years, when the yield is low, the growers do not look for market opportunities, the buyers find them. There are no stable, permanent relationships, although there are regular buyers. If the yields are high, the growers usually take their potato to the wholesale market and they sell it even at a low price.

Most of the buyers are not interested in the cultivar, their only request is that it should be "red-skinned". The most frequent method employed is still direct marketing. About half of the produced potato is marketed this way. Many choose to produce based upon contracting. Accordingly, they produce based on a contract made with a buyer or an integrator organization in advance. With the integrator, a marketing relationship is formed only if a contract has been made in advance, while a high ratio of the buyers purchase the product without a contract. Most of the growers sell the potato in bags and the ratio of those selling in nets is also relatively high. The ratio of those selling in large amounts in containers is lower and only few sell in bulk.

The buyers altogether as market participants do not have a determining role. The most determining factor is the relationship between the price, demand and supply on the wholesale market. The growers will accept the offer of the buyer depending on the fact whether it is better or worse than the price at the wholesale market.

The processing companies have very strict requirements towards their growers, but these are reliable and well-functioning relationships. The small growers do not have a real contact with the retailers or the commercial networks. In general, the retailers do not know the cultivars and the commercial networks have too much sales actions. Therefore, they are willing to buy even the worst quality product, so they can offer it at the lowest price. It has to be mentioned that, for the time being, these networks are receiving the packed, washed, sorted potato, the price of which is much higher than that of the bulk potato, they target the customers with higher requirements. The retailers who want to sell the Hungarian potato abroad are usually from Romania and Ukraine. They know well the cultivars grown in Hungary and buy the high-quality, tasty potato, even if the price of these is a little bit higher.

The participants of the potato industry gain most of their information from the National Potato Product Association, and the most important source of information is the monthly magazine. Some of the growers are also members of the Seed and the Vegetable-Fruit Council. In recent years, the number of the members of grower and marketing associations (GMAs) increased at the highest rate and many have joined the Potato Council. It is probably due to the EU accession that the growers consider it important to belong to some kind of organisation from where they can get information, guidance and assistance to solve their problems in growing and marketing.

Extension experts and educational institutions

These play only a peripheral role in the potato industry. They have few contacts with potato growers and hardly meet with their representatives. The growers obtain limited information from these circles, and usually do not think of asking these institutions for help in solving their problems.

Conclusion

The profitability of the potato production is not uniform, in general early and seed potatoes have a higher profitability than potatoes for winter storage. The profitability is satisfactory above a 20 t/ha yield under favorable price conditions. Under the more stable market situation which presented itself after our EU accession, very high prices cannot be expected. Therefore, the level of production should be improved (to 30-40 t/ha) in such a way that the prime costs are reduced and the quality is improved (Kruppa, 2004). The profitability of the potato industry is higher than that of other field crops but lower than that of the fruit crops. Profitability is highly dependent upon the cultivar. The profitability of the widely grown Dutch cultivars is around the average. However, some farms can obtain higher profits than the average.

Starting from the small farms and ending with the larger farms, the rate of early and summer potatoes falls, while the rate of potatoes with a longer growing period, suitable for winter storage, increases. Almost all of the largest growers grow seed potatoes and are mainly specialized in long storage and autumn harvesting.

Based on the major production data, Hungarian potato production can be successful in the long run, although it has significant disadvantages compared to the European yields per hectare, which enable more favorable costs. However, in order to achieve success, competitiveness should be improved not only through quality products, but also by improving the efficacy of the distribution system and by modernizing the applied production technology.

References

FAO(2005):http://faostat.fao.org/faostat/form?collection=FS.CropsAndProducts&Dmain=FS&servlet=1&hasbulk=&version=ext&language=EN

Kruppa, J. (2004): A magyar burgonyatermesztés lehetőségei. Burgonyatermesztés. V.évf. 3.sz. 11-13p.

Kruppa, J. (2005): A korai burgonya biológiai alapjai, nemesítése, termesztése
Magyarországon, felhasználási lehetőségek. Burgonyatermesztés. VI. évf. 3.sz. 20-27p

KSH (2003): Mezőgazdasági Statisztikai Évkönyv 2002, KSH Budapest.

KSH (2006): http://portal.ksh.hu/pls/ksh/docs/hun/xstadat/tabl2_02_01_03.html

OBTT (2003): http://obtt.hu

Polgár, Zs.(2006). Let's bring back Hungarian potatoes! AgraEconomy of the European
Union. Volume 2006/11., Issue 3. 18-19.

Curriculum vitae – Eszter Zsom

Eszter Zsom - born in a basic crop production region - studied agronomy at Debrecen University, from which she received an MSc in 2001. She then worked and researched for the University of Debrecen, Centre of Agricultural Sciences, at the Department of Crop Production and Applied Ecology 3 years. With a scholarship, she spent one semester at the University of Soil Science in Vienna, at the Department of Crop Production and Crop Breeding, under the supevision of Dr. Kaul, in a research program on soil water management.

She will receive a PhD degree from the University of Debrecen in 2006, on the basis of her work on the influence of irrigation on the yield and quality of potato. The interaction of the influence of irrigation on the yield and quality of potato, as well as the development of irrigated potato production technology, was analyzed for the first time in field trials in this region of Hungary with Hungarian potato varieties.

After she finished her research work, she started to work as a consultant for K+S KALI GmbH. She travels throughout Hungary as a consultant for farmers and agriculture companies who want to develop their potato production technology. She had the results of her research work published in over 20 scientific and conference papers. She is a member of the Hungarian Potato Product Association.

Curriculum vitae – József Kruppa

Dr. József Kruppa is a senior research fellow of the University of Debrecen Department of Crop Production and Applied Ecology. He wrote his doctoral thesis about the impact of lime and dolomite treatments on the fertility of acid sandy soils of the Nyírség region. He has received his PhD degree in 2001, the title of his thesis was: "Results on the breeding and regional production of rye and triticale". He teaches at the University of Debrecen, the University of Mosonmagyaróvár and the Agricultural College of Szarvas. His main research areas in the recent years: development of the production technology of potato, triticale, rye, alfalfa, development of fertilization on acidic sandy soils for potato, triticale, alfalfa, bean, the effect of water supply on the quality of potato tubers.

He has several cultivars of own breeding, 3 alfalfa cultivars, 2 rye cultivars, 1 potato cultivar, 1 triticale cultivar. He has participated in the breeding of 1 common vetch cultivar and 2 millet cultivars.

He is a member of the Hungarian Plant Breeders' Association, Plant Production Committee of the Hungarian Academy of Sciences, board of the National Potato Product Association. In the period between 1993 and 2005, he organized 27 scientific and professional events and he was actively engaged in the organization 36 professional events within the framework of the Dutch-Hungarian potato production program. He has been the editor, author and co-author of several books and co-author of university textbooks.

Company profile – University of Debrecen, department of Crop Production and Applied Ecology

The Department of Crop Production and Applied Ecology - at the University of Debrecen - with its scientific activities reflects the importance of developing and working in harmony with the values and interests of the region. Through its active role in the professional community and through its close connections with the agricultural sector it aims to take a leading role in the improvement of production. Its diverse scientific activities of the Faculty reflect the importance to assist the development of Hungary's eastern region with its long-established agricultural traditions.

The most important fields of study in which the department has carried out research and development cover wheat, sunflower, maize and, during the past few years, also potatoes and ecological crop production.

The laboratories of the university are well-equipped with modern instruments for quality assessment and a special potato quality testing laboratory was founded a few years ago.

The Czech potato crop in the European Union

F. Novák
Central Potato Association of the Czech Republic, Dobrovského 2366, 580 01 Havlíčkův Brod, Czech Republic

Abstract

The potato area significantly decreased in the Czech Republic between 1993 and 2005. The yields do not reach the level of European countries with developed potato production and are unstable. The major cause of fluctuating yields is the variable sum of utilizable rainfalls during vegetation. The proportion of irrigated fields, from which high and stable yields are obtained is small. Long-term storage of high quality trade ware is still a problem. Import of fresh potatoes is annually very variable, depending on achieved yield. The import of potato products has an increasing tendency and is approximately three-times higher than export. The national quota of 33,660 tons of potato starch production is lower than the national consumption and the capacity of starch factories and possibilities of starch potato growers. Delivering large uniform lots of potatoes for processing and fresh consumption after harvest is a strong point. The high volume of the export of table potatoes from the harvest 2005 is to be noted. Good natural conditions for growing of high-quality seed are legislatively supported, so planting of certified seed is only permitted in the best localities. The occurrence of quarantine pests is strictly examined.

Keywords: natural conditions, yields, export, import, Czech Republic

Introduction

Potato growing has a long-year and rich tradition in the territory of the Czech Republic. The production from an important potato area in the first half of last century was particularly used for alcohol production, livestock feeding and relatively high human consumption. From the European level achieved before the World War II, Czech potato growing fell into a deep crisis during the period of socialism. The concentration of areas without powerful mechanization and appropriate storage technology caused low yields and high losses. Between 1961 and 1965, mean hectare yield of 11.5 t/ha was recorded. Technical improvements and a shift of the variety assortment caused a yield increase in the 80ies and the 90ies.

General data

Climatic and soil conditions for potato growing are largely variable in the Czech Republic. Continental and also seaside climate recurrently affects our territory in central Europe. The amount of utilizable rainfalls during vegetation is annually significantly variable. Since the fluctuations are little compensated with small share of irrigated fields, considerable year fluctuations in yields are recorded.

Figure 1 indicates that yielding level in the Czech Republic is still behind the most developed European potato countries, and furthermore it is substantially unstable. For an increase of yields and their stability we consider a necessity to enhance potato farming on irrigated fields and to improve the nutrition level.

The consumption of net nutrients per one hectare of potatoes is substantially lower in our republic than in the countries with the highest European yields. It is not only associated with natural conditions, but also with a worse economic situation of potato growers, given also by substantially lower non-market income than growers from EU-15 have.

In 1993, when the independent Czech Republic was established, the area of 103,000 ha was planted with potatoes in the Czech Republic. From that time on a considerable decrease of the potato area was recorded (Figure 2).

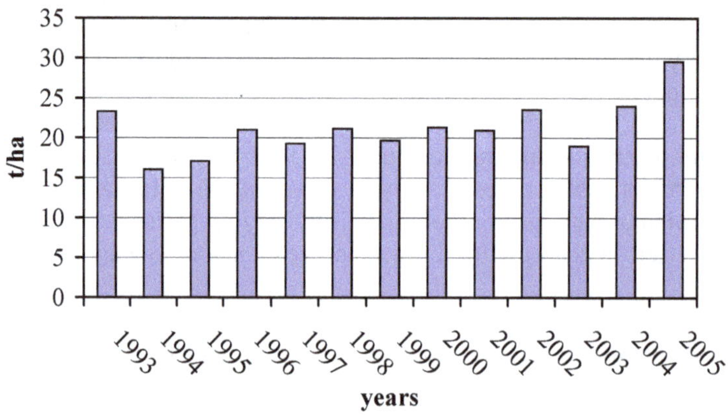

Figure 1. Potato yields in the Czech Republic between 1993 and 2005 (Czech Statistical Office, Prague).

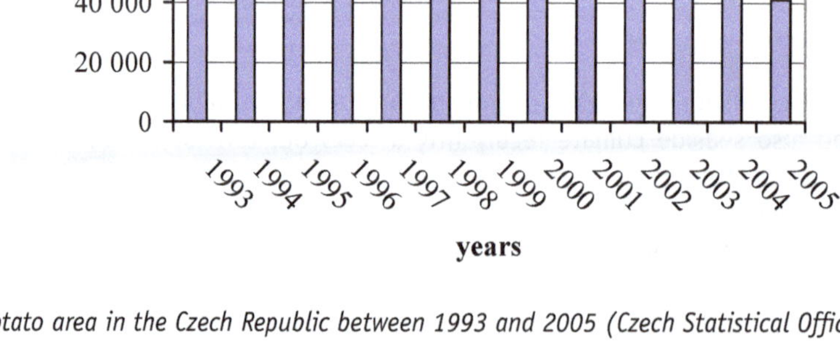

Figure 2. Potato area in the Czech Republic between 1993 and 2005 (Czech Statistical Office, Prague).

The causes of area reduction are:
- stop of potato export to Slovakia in the 90ies (Czech Republic annually exported up to 300 thousand tons of table and seed potatoes into the Slovak Republic);
- animal feeding with potatoes was stopped;
- the quality of production was improved and losses were reduced;
- fresh potato consumption was reduced;
- consumption of potato products, which are predominantly imported, increased;
- yields increased.

Table potatoes

Table potatoes are grown in the Czech Republic in 160 - 650 m above sea level. In lower early potato production regions an important part of the fields are irrigated. Especially, in light sandy soils, when potato crops are covered with non-woven fleece, it is possible to start with the harvest of very early potatoes usually at begin of June. During summer months, high yields are obtained in these regions and production is competitive with the production of neighbouring countries in response to quality and price. In piedmont regions planting into de-stoned soils is applied in a dominant majority of potato areas. Due to the introduction of de-stoning technologies the quality of harvested potato seed, but especially of table potatoes, considerably increased. A relatively large potato area per potato grower is an advantage of our potato production (Table 1).

It allows effective purchasing of demanding technological equipment and deliveries of high uniform lots to the market for fresh consumption and processing. A disadvantage is the insufficient storage capacity for long-term quality maintenance of table potatoes up to the end of the season (May, June).

It has a great influence on gained sales in different periods. Growers are forced to selling for lower prices after harvest. This trend is exemplarily apparent from the development of foreign trade from the harvest of last year (Table 2).

Table 1. Structure of potato growers in the Czech Republic according to sample survey ((Research Institute of Agricultural Economics, Prague).

Potato area (ha)	**Number of enterprises**	**Percent of enterprises**	**Percent of harvested areas**
under 0.5 ha incl.	59	11.09	0.18
over 0,5 up to 5 ha	253	47.56	4.42
over 5 up to 25 ha	110	20.68	10.07
over 25 up to 50 ha	34	6.39	9.91
over 50 up to 100 ha	34	6.39	19.20
over 100 up to 200 ha	29	5.45	29.91
over 200 ha	13	2.44	26.30
Total	532	100.00	100.00

Table 2. Import and export of late table potatoes from July 2005 to March 2006 (t) (Czech Statistical Office, Prague).

Year/Month	05/07	05/08	05/09	05/10	05/11	05/12	06/01	06/02	06/03
Export	1,934	2,076	7,867	9,370	19,569	16,773	4,510	5,094	6,419
Import	809	435	672	598	495	1,038	1,644	6,434	10,635

In contrast to other commodities, a modernization of storage facilities for potatoes was not state-supported before the Czech Republic's entry into the EU, and it is also not state-supported at present. From 2007 on we expect that some subsidies will be provided for this purpose, and that we will catch up with the loss compared to developed potato countries.

Before the Czech Republic entered the EU, the Central Potato Association tried to link necessary contacts and to make conditions for the introduction of our republic into the European market. Our association became a member of EUROPATAT in 1996 and since 2000 we have had the national structure R.U.C.I.P. The trade with the companies in the Czech Republic could be made in accordance with these trading conditions. Our country proclaimed an eradication programme against potato ring rot in 1996. The Czech Republic had and still has one of the strictest inspections of quarantine bacterial diseases (Table 3).

Despite of a more detailed survey, brown rot has not been so far recorded in the Czech Republic. The programme was successful and still before the EU-entry State Phytosanitary Committee admitted equivalence status to the Czech Republic. The Czech Republic is not a risk country as regard as export of table and seed potatoes. For trade with table potatoes within the EU there is not set only one obligatory regulation with minimal qualitative parameters. In our opinion, this fact could complicate future EU trade, since not all new member countries are ready for trade in accordance with R.U.C.I.P. rules, and/or other standards.

Table 3. Frequency of inspections in the Czech Republic and selected EU-countries for presence of bacterial ring rot from the harvest 2003 (European Commission, Food and Veterinary Office, 2004).

Country	Potato seed			Table potatoes		
	Area (ha)	Samples	ha/sample	Area (ha)	Samples	ha/sample
Germany	17,429	11,850	1.5	272,002	4,886	56
France	13,928	7,827	1.8	146,252	882	166
Belgium	2,037	3,731	0.5	57,166	1,204	47
Austria	1,344	662	2.0	18,512	91	203
Poland	6,322	7,685	0.8	778,558	11,818	66
Czech Republic	4,753	3,740	1.3	35,984	1,054	34
EU	115,796	63,160	1.8	2,264,847	31,964	71

Processing potatoes

Before EU-entry we required an annual national quota of 51,000 tons for potato starch production. The requirement corresponded to the capacity of our starch factories, growers of starch potatoes and also our national consumption of potato starch. During pre-entry negotiations a quota of only 33,660 t was assigned to us. Due to this quota we were doomed to the role of constant importers of potato starch. Since production capacities are underutilized, the production becomes more expensive.

At the potato product consumption, imported products are predominant. While the import of table potatoes annually is subjected to strong fluctuations, the import of potato products has a stable tendency (Figure 3).

In the Czech Repulic, no important processing capacity is available and existing smaller ones assert themselves in the competition on a European scale only with difficulties. We have the active balance of trade only for potato crisps; for French fries the balance is deeply passive. Since the growers sell the raw material to the processors for the European prices in our country, they are searching the market for their potatoes in neighbouring countries as well. The future will show, whether or not only fulfilment of qualitative parameters, but also logistic assurance will be in our power and the co-operation over the borders will be successful in this area.

Seed potatoes

Together with the overall reduction in the potato area, the area of seed-potato is also reduced (Table 4). In 2005, 177 potato varieties were submitted to the registration. The Czech varieties were grown on 860 hectares. Good climatic conditions for the production of high-quality potato seed are also supported by legislature in our country. The law prohibits planting of other than certified seed in the most favourable potato production regions. In this way the occurrence of

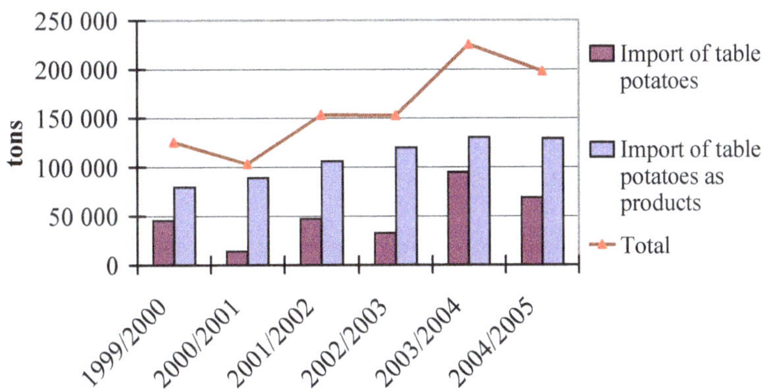

Figure 3. Import of table potatoes and potato products (converted to fresh potatoes) into the Czech Republic within the fiscal years 1999-2004/05 (Czech Statistical Office, Prague).

Table 4. Changes in seed-potato area in selected years in the Czech Republic (Central Institute for Supervising and Testing in Agriculture, Brno).

Year	1993	1994	1997	1998	2000	2001	2003	2004	2005
Area (ha)	11,159	6,559	6,542	5,291	5,706	5,262	4,754	5,030	4,424

virus disease sources is limited. This measure with good work of seed-potato growers allows achieving of really high-quality production.

After the Czech Republic's entry into the EU our balance of foreign trade with seed-potatoes has considerably changed. Previously, our country was only an importing country, from the harvest 2004 the balance has improved, and from the harvest 2005 the balance has become active. We are sure that in near future our favourable geographic position between breeding centres in the Netherlands and Germany and prospective customers of potato seed in eastern and south-eastern Europe will be more utilized than at present. We expect speeding up of this trend due to an increase of transport cost. Hopefully, several Czech varieties will be introduced to the EU-market in the future.

Conclusion

The Czech potato growing does not threat potato growing in neighbouring countries in regard to the volume and competition ability. In the future, there will be problems due to low quota for potato starch production and split production capacities for potato processing into products. Long-term data indicate the fact that the fastest reduction in potato areas in Europe occurs in the countries with less-developed processing industry. When the conditions of the Czech Republic will be improved for a long-term good storage of potatoes, the share of irrigated fields and total yields will be increased, and potato production could be competitive in terms of free trade. Hopefully, the non-market income of our potato growers and growers of EU-15 will be equal in the nearest future. This will contribute to the balancing of competition terms. The Czech potato growing is not threatening from a quarantine point of view. There are aspirations for production and supply of high-quality potato seed to the interested people in Europe based on a strict control of propagating crops by the Czech supervising institute.

Curriculum vitae – František Novák

František Novák studied agronomy at Agricultural University in Prague, and graduated in 1969. He worked for ten years as an agronomist in two united agricultural cooperatives specialized in seed-potato production. Between 1980 and 1992 he worked at the Potato Research Institute in Havlíčkův Brod. At first, in the department of genetics, later in the department of biotechnologies. He was the leader of this department, when utilization of tissue cultures was introduced into maintenance breeding of potato in the Czech Republic in large scale. He was involved in the research of extreme intolerance to Potato leafroll virus and took PhD. degree in connection with this work in 1997. In 1992 he was a member of the Federal Assembly of the

Czech and Slovak Federal Republic. Between 1993 and 1996 he was the secretary of the Central Potato Association of the Czech Republic. Since 1996 till now he has been the president of the Association. Between 1996 and 1998 he was a member of the Chamber of Deputies of Parliament of the Czech Republic and a vice-president of its Agricultural Committee. Since 1998 he has been the director of the company AGRORADOST s.r.o. Havlíčkův Brod. The company propagates seed-potato and was the largest exporter of potato seed in 2005.

Company profile – Central Potato Association of the Czech Republic

The Central Potato Association of the Czech Republic (ÚBS ČR) was founded in 1993 as a voluntary association of potato growers, processors and traders. It is divided in the sections of growers, processors and traders. Each section directly elects its chairman (who is also a vice-president of the Association) and its deputies for Association Board. The president is elected by the General Assembly. The director of the Potato Research Institute is a member of the Board. The Board has the equal representation of sellers (potato growers) and buyers (processors and traders). The members represent the whole vertical of branches from research through breeding of new varieties, propagation of seed potatoes, growing of marketable potatoes, trade with certified seed potatoes and market goods to industrial processing of potatoes. Due to its memberships and structure of the Board the Association can make and enforce its standpoints from the view of the whole potato industry. It actively participates in the preparation of new regulations affecting branches and informs other members about the regulations. It has also a consulting activity, which part is *e.g.* publishing of the journal Bramborářství (Potato Growing), publishing of the Catalogue of Potato Varieties Registered in the Czech Republic, testing of varieties and publishing of List of Potato Varieties Recommended for Starch Production *etc.* Due to its own information systems it mediates information from inland and abroad as well. Since 1996 ÚBS ČR has been a member of EUROPATAT.

Actual state and future trends of the Romanian potato industry

S.C. Chiru, Gh. Olteanu and Elena Laura Dima
National Institute of Research and Development for Potato and Sugar Beet (NIRDPSB) P. O. Box 500470, Brasov, Fundaturii st. 2, Romania

Abstract

Romania is preparing to become a member of the European Union in 2007. The potato industry must be adapted to the European competitive system. According to the importance of the potato crop (area and total yield), the seed production, and the market demands for varieties and consumption (fresh consumption will still be on the first place), a current analysis of the Romanian potato industry and future tendency based on a SWOT interpretation (strength, weakness, opportunity, threat) is carried out.

Keywords: statistic data, varieties, seed production, SWOT analyses, EU accession

Introduction

Romania is situated in the South-East part of Europe, at the intersection of principal communication axes between North-South and East-West. With a total area of 238,391 km^2, almost as large as the United Kingdom (244,100 km^2), Romania would become the ninth largest country in an enlarged European Union with 27 members states. It will represent 5.4% of the area and 4.7% of the EU population (MAFRD, 2006).

The Romanian agriculture, although with a high natural potential, is still far away from a compatible level to EU structures. From the 23.8 million hectares of total area, agriculture amounts to 14.8 million hectares. The overall crop production covers 8.88 million hectares (2004), of which cereals account for 62%, oleaginous plants for 15%, and potatoes for 3.2% (Salasan, 2003).

Actual state

Potato production has a relatively long tradition in the Romanian agriculture, first references being made in Transylvania in the 18th century. From that time on, crop importance increased and potatoes are now considered as the "second bread" of Romania. Excepting the years between 1970-1990, when the potato production was concentrated on an industrial level, both in anterior periods and in the present situation small plots prevail.

The EU joining of Romania in 2007 imposes a new production system for potato growing (Olteanu, 1997, Chiru, 2006):

- Changing the mentality of stakeholders;
- Increasing the surfaces of agricultural exploitation;

- Constitution of producers associations;
- Orientation in terms of the market (consumers requirements);
- Monitoring the potato chain (traceability);
- Sustainable crop production (environmental friendly);
- Diversification of potato processing.

In Romania, the potato is considered as a strategical food, being a component of the food security system. Energy, protein and dry matter production per square unit is the same as that produced by the main crops (wheat and maize) (Chiru and Olteanu, 2004). Considering the annual capita consumption of potatoes (95.1 kg), Romania is the third maxiumum country of Europe, whereby the range is between 39 and 127 kg (Figure 1).

Production and surfaces evolution

In the last 50 years (1950-2005), the total cultivationof potatoes covered 250,000-316,000 ha (Figure 2).

In 2005, the area of 285,000 ha placed Romania on the second rank in Europe after Poland (Figure 3).

The average yield of potatoes accounts for 10-14 to/ha, still much lower than the average yield in the EU with 36-40 to/ha (Figures 4 and 5). For 2007, Romania has to follow the EU tendency in the sense of area reduction and yield increase (Figure 5).

Biological resources (variety, seed)

Quality and quantity characteristics of Romanian and foreign varieties concerning growing years and locations with different soil typs and climatic conditions were investigated. Between 1978 and1985, the native varieties Semenic, Super, Colina and Sucevita were cultivated on 40% of the total potato area. Between 1990 and 2004, main varieties were the Dutch ones Sante

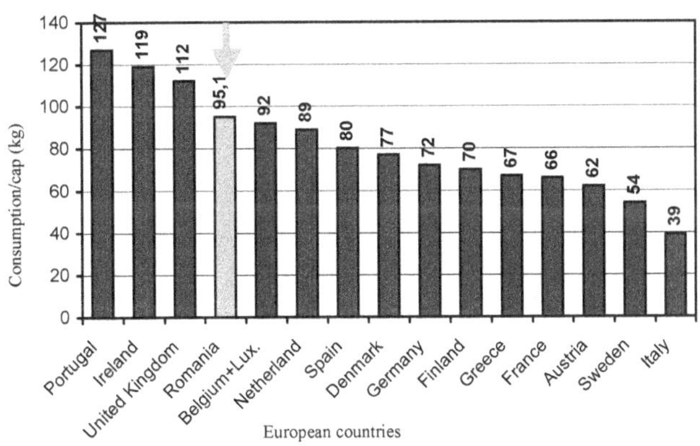

Figure 1.The rank of Romania in Europe by potato consumption (Cap/Yr) (FAO Databases, FAO 2006).

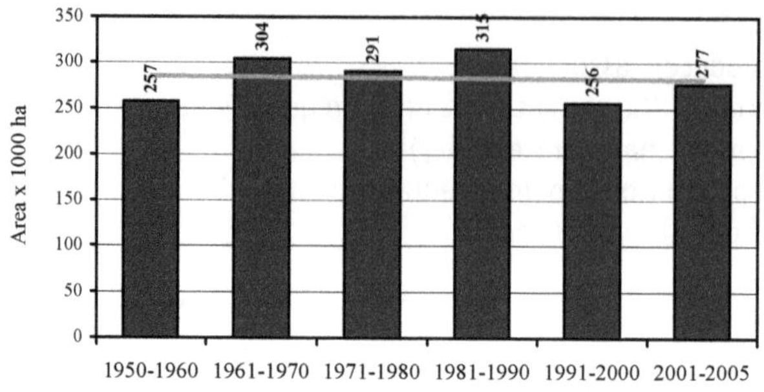

Figure 2. Potato area in Romania between 1950-2005 (FAO Databases, FAO 2006).

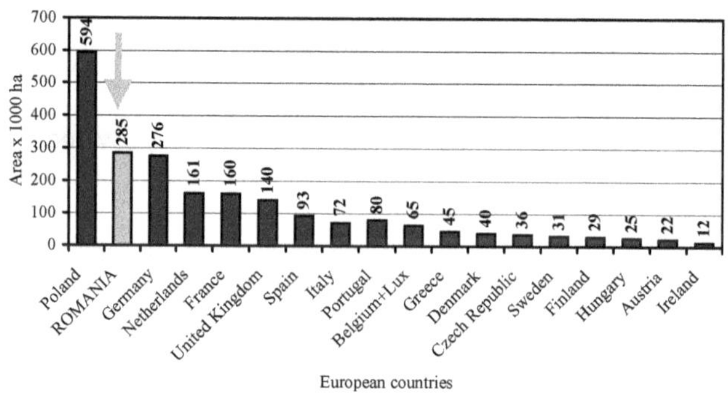

Figure 3. The rank of Romania in Europe by the potato area (FAO Databases, FAO 2006).

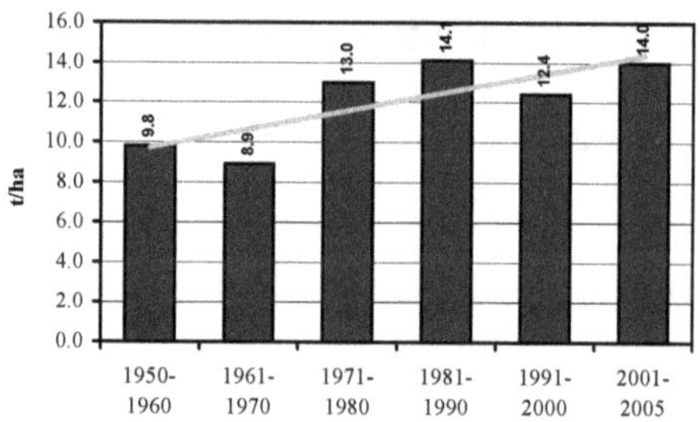

Figure 4. Evolution of potato average yield in Romania between 1950-2005 (FAO Databases, FAO 2006).

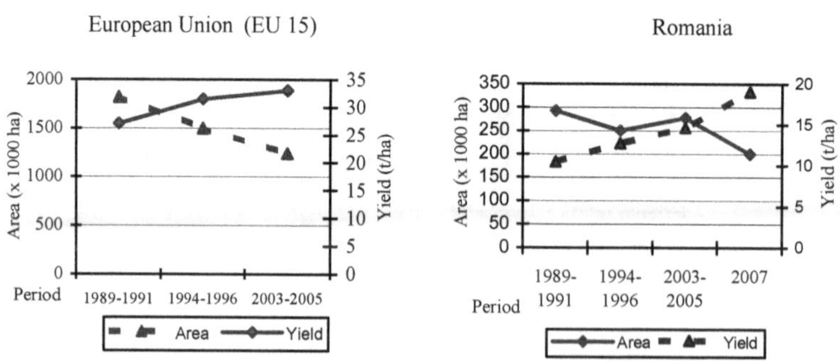

Figure 5. Trends in the evolution of potato area and yield in EU and Romania Targets for 2007 (FAO Databases, FAO 2006).

(36%), Desiree (31%) and Ostara (11%). These varieties were cultivated in Romania during the last 20 years (Bozesan, 2002, Bozesan *et al*, 2004). In 2005, the National Catalogue of varieties included 80 potato varieties (MAFRD, 2005). From these, 30 varieties issues from EU countries. The national seed production system incudes 20 varieties (Table 1).

After 1990, the seed potato area decreased in our country. The landowner Law no.18/1991 produced a dramatical change regarding the agriculture proprieties, with influence on the national seed potato production. In 1990, 25,000 ha were cultivated with seed potatoes, but the area was reduced to 6,400 ha in 1999, and to 1,800 ha in 2005 (Table 2). Small growers (farmers) do not strictly respect the required seed potato treatments, generating a large proportion of non-certified seed (10,800 t in 2004).

Table 1. Variety structure of the national system for seed potato production in 2005 (NFP, 2005).

County/Varieties	BC	BT	BV	CV	HG	HD	IS	NT	SB	SV	Total
Romanian varieties > 1 ha											
Armonia	-	-	-	4	-	-	-	-	-	-	4
Christian	-	-	8	-	-	-	-	-	-	-	8
Luiza	-	-	-	3	-	-	-	-	-	-	3
Mikel	-	-	-	12	-	-	-	-	-	-	12
Milenium	-	-		7	-	-	-	-	-	-	7
Redsec	-	-	-	15	-	-	-	-	-	6	21
Roclas	-	-	30	15	-	-	-	-	-	-	45
Speranta	-	-	-	6	-	-	-	-	-	-	6
Star	-	-	-	5	-	-	-	-	-	-	5
TOTAL	-	-	38	63	0	0	0	0	0	6	69
Foreign varieties > 20 ha											
Agata	-	-	18	6	5	-	-	-	-	-	29
Desiree	2	18	68	14	91	6	23	104	5	32	363
Hermes	-	3	19	21	13	-	-	-	-	8	64
Impala	-	-	41	40	6	-	-	2	17	2	108
Kondor	2	-	44	39	-	20	24	2	-	69	200
Kuroda	-	-	-	20	-	-	-	-	-	-	20
Laura	-	-	-	5	-	-	-	-	34	-	39
Ostara	-	-	2	18	-	-	-	8	-	-	28
Riviera	-	-	9	23	-	-	-	-	-	-	32
Rosara	-	-	13	-	-	-	-	-	-	10	23
Sante	-	4	228	4	188	19	-	3	8	122	576
TOTAL	4	25	442	190	303	45	47	119	64	243	1,482

Legend: BC-Bacau, BT-Botosani, BV-Brasov, CV-Covasna, HG-Harghita, HD-Hunedoara, IS-Iasi,. NT-Neamt, SB- Sibiu, SV-Suceava

Table 2. Seed potato production from 1999 to 2005 (area: 1000 ha; yield: Mt) (TISCPM, 2005).

Year	M.U.	Seed potato area/Planned production	Total	Declassified	Biological categories					Rejected seed
					Prebase	SE	E	Cl.A	Cl.B	
1999	ha	6.4	5.4	0	-	0.4	1.2	1.9	1.9	1.05
	t	87.2	86.0	0	-	6.6	18.6	30.3	30.5	1.2
2000	ha	4.9	3.5	0.5	-	0.08	0.7	1.4	1.4	1.4
	t	51.7	49.3	6.8	-	0.9	9.9	19.0	19.5	2.4
2001	ha	5.2	4.1	0.5	0.003	0.08	0.6	1.3	2.2	1.04
	t	68.5	65.8	7.8	0.06	1.7	1.8	26.5	35.8	2.7
2002	ha	3.4	2.8	0.4	0.001	0.07	0.16	1.3	1.3	0.6
	t	52.5	51.6	7.5	0.02	1.3	3.2	24.2	22.9	0.9
2003	ha	-	2.8	0.05	0.04	0.14	0.2	1.03	1.4	-
	t	-	44.9	1.1	0.7	2.2	3.4	16.5	22.1	-
2004	ha	3.2	2.9	0.2	0.04	0.2	0.5	1.22	1.03	0.2
	t	50.0	43.4	4.2	0.4	2.05	7.1	17.5	16.4	6.6
2005	ha	1.8	-	-	0.065	0.08	0.2	0.96	0.5	-
	t	28.5	-	-	0.9	1.12	3.05	15.4	8.02	-

Analyzing the Romanian potato market and looking to consumers' tendency, the potato council (stakeholders) estimates a production of 3,590 Mt in 2006 to be distributed upon the different utilization categories (Table 3).

It is to be noticed that a low amount of potatoes is used for processing (2%), because of few existing processing plants.

Table 3. Potato production and its utilization (proposal for 2006) (RPC, 2006).

Specification	Area (1000 ha)	Quantity (Mt)
A. Potato for human fresh consumption	232	3,480
Very early and early potatoes	10	150
Medium late potatoes	20	300
Late potatoes	202	3,030
B. Forage potatoes	-	500
C. Potatoes for processing	3	60
D. Seed potatoes	2.5	50
Total	237.5	3,590

SWOT analysis

Based on the situation presented above, the principal strengths, weaknesses, opportunities and threats (SWOT analysis) have been identified for the potato industry chain (Olteanu *et al.*, 1997).

Strengths:
- Long tradition on potato crop.
- Favorable land and climatic resources.
- Agronomy/research/extension support available.
- Suitable potato crop zoning.
- Good local market potential.
- Favorable conditions for seed potato crop.
- Adequate potato varieties.

Weaknesses:
- Agricultural exploitation with small acreage.
- Relative high production costs.
- High risks (especially weather).
- Low technical and financial means of farmers.
- Small-sized farms with a reduced professional knowledge.
- Ageing structure of farmers.
- Small number of up to date farms.

Opportunities:
- New varieties (better yield/quality).
- Existing promoting organizations (Product Council, National Federation of Potato).
- Demand of potatoes for the CIS market (Commonwealth of Independent States) and Turkey.
- Potato export possibilities on eastern markets.

Threats:
- Global climate change.
- Rising cost of land and utilities.
- Instability of the potato market.
- Lack of guaranteed minimum prices.
- Insufficient development of potato processing.
- Overproduction in western European countries.
- Competition of economically advanced countries.

Conclusions

In Romania, the potato industry will maintain its great economical and social importance also after an EU accession in 2007. The main challenge will be the reorganization of the whole potato chain to enter the European level. This will impose enlargement of the agricultural exploitation, new organizational structures of the potato chain, and initiation and professional preparation of those who are implicated in the potato industry. A selection of the most adapted varieties for Romanian conditions will be accompanied by a modernization of the seed potato production with view to provide internal and export market demands.

References

Bozesan (2002). Realizari si perspective privind genetica si ameliorarea cartofului (Achievements and perspectives of potato genetics and breeding). Anale ICDCSZ Brasov, XXIX: 16-36.

Bozesan, I., S.C. Chiru and C. Draica (2004). Performantele noilor soiuri de cartof create de Institutul de Cercetare-Dezvoltare pentru Cartof si Sfecla de Zahar (Performance of new potato varieties bred at Institute of Research and Development for Potato and Sugar Beet). Proceedings of EAPR Agronomy Section Meeting, Mamaia, Romania, June 23-27, 2004. Anale ICDCSZ Brasov, XXXI: 35-41.

Chiru, S.C. (2006). Proiect managerial de dezvoltare al Institutului National de Cercetare-Dezvoltare pentru Cartof si Sfecla de Zahar Brasov (The managerial development plan of the National Institute of Research and Development for Potato and Sugar Beet Brasov). Personal comunication: pp. 37.

Chiru, S.C. and Gh. Olteanu (2004). Research and development priorities in potato field with to sustainable development of Romanian agriculture. Proceedings of EAPR Agronomy Section Meeting, Mamaia, Romania, June 23-27, 2004. Anale ICDCSZ Brasov, XXXI: 2-15.

FAO (2006). Food and Agriculture Organization of the United Nations, FAO Statistical Databases FAOSTAT. http://www.fao.org.

MSFRD (2005). Catalogul oficial al soiurilor de plante de cultura din Romania (Oficial Catalog of crop varieties in Romania). Ministry of Agriculture, Forest and Rural Development: 34-35.

MSFRD (2006). Planul national strategic, 2007-2013 (National strategic plan, 2007-2013). Ministry of Agriculture, Forest and Rural Development: pp. 76.

NFP (2005). Variety study of national sistem for seed potato production. National Federation of Potato. Personal communication: pp. 8.

Olteanu, Gh. (1997). Agricultura de precizie, un nou concept in cercetare si managementul agricol (Precision farming, a new concept in research and agriculture management). Anale ICDCSZ Brasov, XXIV: 85-92.

Olteanu, Gh., C. Draica, Gh. Pamfil, B. Plamadeala and I. Bozesan (1997). Prioritati in cercetarea stiintifica si dezvoltarea tehnologica la cartof (Priorities in potato scientific research and tehnological development). Anale ICDCSZ Brasov, XXIV: 30-36.

RPC (2006). Potato production structure for 2006. Roumanian Potato Council. Personal communication: pp. 2.

Salasan, C. (2003). Problem solving group - Agricultural extension method. PhD Thesis. Banat's University of Agricultural Sciences and Veterinary Medicine Timisoara: pp. 263.

TISCPM (2005). Seed potato production - 1999-2005 (Statistics). The Territorial Inspectorate for Seed Control and Planting Material. Personal communication: pp. 5.

Curriculum vitae – Sorin Claudian Chiru

Sorin Claudian C. Chiru, born on 22th September, 1953 in Braşov, a well-known Romanian potato region, studied Agronomy at the Bucharest University of Agricultural Sciences and Veterinary Medicine and received an Agricultural engineer degree in 1978 and a PhD degree (for a work on hybrid small populations used in potato breeding) in 1998. He worked in the breeding and seed potato production group at the Potato Research Institute Braşov (1979-1994), and for 9 years (1995-2003) in two international chemical companies (American Cyaamid and BASF).

Since 2004, Dr. Chiru acts as the General Director of the National Institute of R&D for Potato and Sugar Beet (NIRDPSB) in Braşov. He is the breeder of 9 potato varieties and the author of over 60 scientific papers, and is leading 3 national research projects at potatoes. He has had personal trainings in potato topics in the Netherlands, in Germany, Belgium and France. He is a member of some professional organizations, and from 2005 on he is the President of the European Association of Potato Research (EAPR).

Company profile – National Institute of R&D for Potato and Sugar Beet (NIRDPSB) Braşov

The National Institute of Research and Development for Potato and Sugar Beet (NIRDPSB), founded in 2006 by a reorganization of the former R&D Institute for Potato and Sugar Beet is continuer of a long tradition (57 years) in research and development of two major crops in the Romanian agriculture. It is a top-level institution depending on the Academy of Agricultural and Forestry Sciences (AAFS) and the Ministry of Agricultural, Forestry and Rural Development (MAFRD). It coordinates the national research and development programmes in the field of potatoes, sugar beet and medicinal plant production, storage and marketing.

The institute carries out its research activities in laboratories, greenhouses, experimental fields, and on its own pilot farms, or in cooperation with others national and international units (research organizations, privates companies *etc.*).
The current main research topics are:

- breeding of new potato, sugar beet and medicinal plant varieties, adapted to drought and temperature stress, and tolerant to pests;
- improving integrated, environmental friendly technologies for crops used in the rotation of upper land system;
- improving the methods for basic seed production;
- precision farming for potatoes and sugar beet;
- diversified methods for potato marketing.

NIRDPSB has divided its activity into 3 units:

1. Breeding and basic seed production.
2. Technologies and precision farming.
3. Marketing.

The institute will organize the XVII Triennial Conference of the EAPR in Braşov in 2008.

Potato in Estonia: Production and research

K. Kotkas
Research Centre EVIKA Tallinn University of Technology, Teaduse 6a, 75501Saku, Harjumaa, Estonia

Abstract

The intensive cultivation of potato in Estonia began in 1850. The planted area of potato was the largest in 1952 - 104,000 ha and the production was the highest in 1968 - 1.521.900 t. For now the officially registered area of potato is ca 16,000 ha and total production is around 300,000 t. Potato varieties are bred in Estonia for more than 100 years. There are 40 national varieties currently registered in Estonia. In the last century potato was used in spirit, starch and chips' production, and a considerable amount of seed and consumption potato was exported. Now mainly table and seed potato is produced for local consumption. In Estonia 4 institutions in potato research are involved.

Keywords: research in Estonia, seed potato, production data

Introduction

Potato is considered the second bread in Estonia - it is very popular component in peoples' diet. However, potato is a relatively young crop in Estonia. Potato was probably brought to Estonia from Russia in 1730-1750. The intensive cultivation of potato in Estonia began in 1870-1880, thanks to the fast development of spirit production. At the beginning of 19th century (1900-1913) the planted area of potato in Estonia was 70,000 to 80,000 ha and the yield 10-12 t/ha. Potato took about 25% of the total area of arable land. About 26.4% of the production was used for human consumption, 25.7% as seed potato, 22.4% in spirit production and 7.0% was exported.

Starting from 1934 the planted area of potato has constantly been over 70,000 ha. When the need for fresh potato declined producers started to look for channels to market the seed potato instead. Estonian seed potato was exported to many central- and southern-European countries but also outside of Europe already in 1932. Consumption potato has been exported from Estonia to Soviet Union, Germany, UK, USA, France, Norway, Latvia; seed potato to Italy, Spain, Portugal, Switzerland, South-America.

The largest planted areas were between 1951-1962, reaching from 93,100 ha (1955) to 104.800 (1952). Since 1963 (89,000 ha) the planted area started to decline gradually and for 1991 potato was grown on 52,300 ha in Estonia. In the mid-nineties the potato area stabilized for some period and was about 35,000 ha but before the new millennium a new decrease began. The climatic conditions vary from moderately favorable to unfavorable from year to year, total production has been 300,000-500,000 t.

Present situation

Relative share of agriculture in GDP and employment

Agriculture is an important economic sector that supplies the population with domestic foodstuffs and ensures employment in rural areas and landscape maintenance. Estonian agricultural policy changed sharply at the beginning of the 1990s: from highly subsidised agricultural production to virtually unsubsidised production with a free price formation. As a result, employment in the sector decreased, as did the relative share of the added value of agricultural products in gross domestic product (GDP) (Table 1). Due to structural changes in agricultural holdings and their re-orientation in the new economic conditions, nearly 25% of arable land was temporarily disused. The relative share of agriculture in GDP and employment are close to the respective indicators of other Baltic countries.

In 2004 as Table 1 shows, the relative share of agriculture in GDP has decreased 0.5-fold since 2000 (www.ref 5, 2004). The added value for agricultural products in 2004 was 3.4 billion crowns in current prices, increasing 17% compared to year 2003. The share of agriculture in the gross domestic product (GDP) was 2.7% - showing a decrease of 20% compared to the year 2000. The relative share of agriculture in the employment of rural population has decreased (5.0% in 2000, 4.1% in 2004) (www.ref 7, 2004). In 1992, 51% of rural working-age population was employed in agriculture, but in 2000 only 18%.

Land resource and use and structure of farming system

The use of the main agricultural production resource, agricultural land, which decreased at the beginning of the 1990s, stabilised by the turn of the century. The land cadastre contained records on 686,267 ha of arable and garden land and 170,682 ha of natural grassland as of the end of 2001 (a total of 856,949 ha of agricultural land is thus registered). Considering that the area of agricultural land before land reform was 1.12m ha of arable land and 240,000 ha of natural grassland, the registration percentages are 61.3% and 71%, respectively (www.ref 3, 2001).

According to the preliminary data of the agricultural census carried out in 2001, the area under field crops was 745,000 ha. According to the expert assessment of specialists of the Ministry of Agriculture, the area under grains was 270,000 ha and the area under leguminous vegetables was 3500 ha in all production units collectively (www.ref 1, 2003). Perennial grass plants were grown on 390,000 ha, rape and turnip rape on 27,500 ha, potatoes on 22,000 ha, open field vegetables on 3300 ha and fibre and oil flax on *ca* 100 ha. There were 341,000 ha of unused arable land and 33,600 ha lay fallow (both black fallow and green-manured fallow).

Table 1. Relative share of agriculture in GDP and employment in 2000-2004 (www.ref 5, 7, 2004).

Year	2000	2001	2002	2003	2004
Percentage of GDP	3.2	3.0	2.6	2.5	2.7
Persons employed in agriculture and hunting, '000	31.5	29.0	30.1	25.9	24.2
Employment rate, %	5.0	5.0	5.1	4.4	4.1

The main reason for the decrease in areas under field crops is the slow pace of land reform and agricultural reform, economic problems resulting from the low income of agricultural holdings and farms, lack of current assets, lack of cultivation machinery and difficulties in receiving low-interest credit.

The structure of farming system

According to the preliminary data of the agricultural census, there were 85.300 agricultural farms and 176.400 households in Estonia. An agricultural farm is a production unit that has at least one hectare of agricultural or forest land, at least 0.3 ha under fish ponds or where agricultural products are produced chiefly for commercial purposes.

According to 2003 structural study data, there were around 36.800 farms and 795.600 ha of agricultural territory (Table 2). 70% of farms had less than 10 ha of agricultural land and owned 12% of it (www.ref 6, 2003). Most (66%) agricultural land belongs to farms with over 50 ha of agricultural land. There was an average of 244 ha of agricultural land in the possession of such farms.

The number of households where potato is grown is decreasing but the acreages of the production farms are increasing (Table 3) (www.ref 4, 2005).

Table 2. The distribution of farms and arable land in their use by size in 2003 (www.ref 6, 2003).

	Total	**Size of production area (ha)**						
		<5	**5-<10**	**10-<20**	**20-<30**	**30-<50**	**50-<100**	**>100**
Number of farms	36,792	18,671	7,264	5,347	1,889	1,482	1,051	1,090
Arable land, ha	795,600	45,929	50,546	74,004	45,493	56,422	71,704	451,542

Table 3. The distribution of potato productions farms in their size (www.ref 4, 2005).

Year	**Total**	**Size of production area (ha)**						
		>0-<1	**1-<2**	**2-<5**	**5-<10**	**10-<20**	**20-<30**	**>30**
2001	44,213	41,075	2,292	660	101	85		
2003	24,961	22,454	1,789	506	124	52	20	17
2005	16,430	14,803	993	441	112	46	11	24

The potato production area and yield

The main area of plant production in Estonia is feed production, the main crops being grasslands and grain for fodder. The area under potatoes is decreasing year by year (Table 4). Data presented in the table are collected from the registered farms. Pursuant to the Plant Protection Act that took effect on 1 May 2004, the plant health register also must list persons who market consumer grade potatoes or grow the crop for their own use on more than one-half hectare. Therefore the real potato production area together with growers smaller than 0.5 ha was as following: 2001:22,000 ha, 2003:14,300 ha, 2005:15,800 ha. One reason for the decrease of potato production area is the unchanging marketing price of potatoes while the expenses have been gradually increased over the last years.

Simultaneously with the decline of production area the production itself declined (Figure 1). The total yield was the lowest in 2004 - 178.900 t that was 36.6% less than in 2003 (244,400 t). In 2005 the planted area of potato decreased but the total production increased. The average yield has been 15-16 t/ha over the past years, the yields of the biggest growers reached up to 35-45 t/ha.

Seed potato production

The production area of seed potato has been below the acreages that is needed for proper seed renewal for many years and has decreased year by year: in 2001 - 110 ha, in 2002 - 103.6 ha and in 2003 - 83.9 ha. Since 2004 the seed potato area has started to increase. According to

Table 4. The distribution of potato production in their size (www.ref 4, 2005).

Year	Total	Size of production area (ha)						
		>0-<1	1-<2	2-<5	5-<10	10-<20	20-<30	>30
2001	17,153	10,268	2,582	1,721	662	1,921		
2003	12,554	6,338	1,956	1,339	795	678	473	975
2005	9,675	4,455	1,196	1,229	730	600	253	1,212

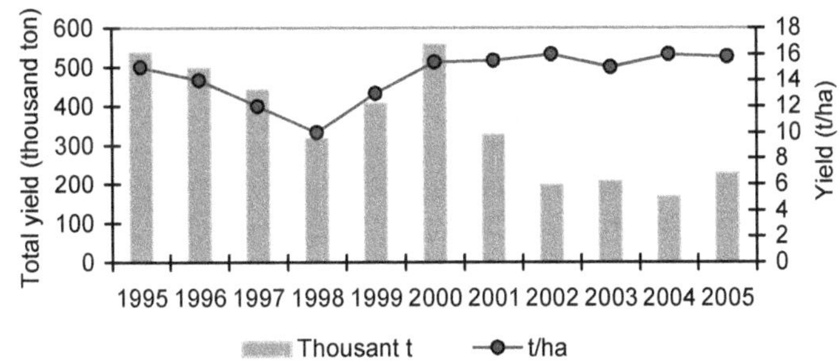

Figure 1. The yield and total yield in 1995-2005 (www.ref 4, 2005).

the data of the Plant Protection Inspectorate the area of approved/certified potato production were 113.3 ha in 2004 (www.ref 2, 2005). Considering the basic yield of seed fraction being 10 t/ha we might be able to produce total 1130 t of seed potato that would supply the seed material for 400-450 ha only. Unfortunately, the growing area of seed potatoes has been insufficient to cover the seed demand for many years and the situation did not change for the better in last year. In 2005, the approbation was applied for 207 ha of seed potato, out of which more than 20 ha (ca 10%) did not qualify and was left uncertified. In total 186.6 ha of potato fields were certified.

In 2005 according to the data of the Plant Protection Inspectorate, field approbation of seed potatoes was carried out for 39 varieties. Among the approved area Estonian varieties ('Anti', 'Ants', 'Maret', 'Piret', 'Sarme') were grown on 16 ha (8.6%). The area seems rather small compared to the rest of the varieties but it should be taken into account that the main approved category of local varieties is SE2 and SE3 that cannot be grown on extensive acreages. Only 4.8 ha (30%) of Estonian varieties were approved as E3 or A. The main approval categories for foreign varieties are either E3 or A since E2 or E3 category seed is imported. Therefore the approved acreages are larger. On the basis of the areas of approved seed fields at last years it may be said that the following varieties dominate: 'Van Gogh', 'Anti, 'Ants', 'Berber', 'Vineta', 'Piret', 'Asterix', 'Folva' and 'Agrie Zeltene'.

Under the conditions of liberal market economy there is no reliable information currently available about the varieties grown in Estonia overall. We also lack the information of the household varieties. In 2005, the Plant Production Inspectorate carried out a monitoring of potato bacterial diseases on 1945 ha *i.e.*12% of total potato area. During the monitoring 94 varieties + fields with mixed varieties and with no specific varieties were registered.

Out of all detected varieties, 10 (33.2%) were the ones bred at Jõgeva Plant Breeding Institute that were grown on 642.2 ha. Out of all varieties registered during that monitoring program 'Ants' was the variety grown on the larges area (22.4%). From Estonian varieties also 'Anti' (6.5%), 'Ando' (2.5%) and 'Piret' (0.8%) were grown on rather large areas. From foreign varieties 'Folva' (13.4%), 'Van Gogh' (10.5%) and 'Agrie Zeltene' (7.3%) were the three most popular varieties. These three varieties made 1/3 of the monitored area, similarly to the domestic varieties. On the last 1/3 of the studied area the rest of 81 varieties and potatoes of no particular variety were grown.

The scope of potato varieties grown has greatly increased during the recent years. Varieties were selected for the yield and for technological qualities. The use of potato is taken into account when selecting the potato variety:

- Table potato - varieties with a longer rest period, and a more stable quality.
- French fries potato - varieties with oblong or oval tubers that absorb oil and maintain their yellow colour. Catering enterprises use Dutch and Belgian varieties grown and imported from Poland. A predominant variety is the German Agria.
- Potato chip production has been terminated in Estonia because potatoes grown in Estonia had large dark spots on the surface and inside.
- Purée potatoes are imported from Sweden and packaged in AS Lekto. Snacks from potato meal imported from France are made in Ääsmäe.

Research on potato

In Estonia 4 institutions are involved in potato research: Jõgeva Plant Breeding Institute, Estonian Research Institute of Agriculture, Estonian University of Life Sciences and Research Centre EVIKA Tallinn University of Technology.

1. Jõgeva Plant Breeding Institute

The plant breeding institution at Jõgeva was founded in 1920. Jõgeva PBI as a single institution in Estonia, involved in plant breeding research and seed production of arable crops, guarantees the economic sovereignty of Estonian plant husbandry.

The main areas of activities of the Institute are plant breeding of agricultural crops and maintenance breeding, production of breeder's seed, production and marketing of certified seed, basic research on agro technical aspects of agricultural crops and maintenance of plant genetic resources in the gene bank.

The aim of breeding programs is to obtain high-yielding varieties characterized by genotype x environment stability of traits. High level of resistance to plant diseases is essential from the aspect of environment protection. All aspects of crop quality have become more important - varieties for food consumption must have high quality characteristics.

Potato breeding is carried out for different purposes (salad, French fries, cooking, hobby growers). During the last decade following potato varieties were developed by the Jõgeva PBI: 'Anti' (1997), 'Juku' (1997), 'Ants' (1992), 'Piret' (2000), 'Maret' (2003). The varieties of Jõgeva PBI have been successful on the domestic market.

Estonia has adjusted it legislation concerning plant breeding and seed production and has become a member of international conventions and organizations such as UPOV, ISTA, OECD *etc.* regulating the utilization of cultivars and seed trade, occurrence of long-term traditions, which make the local cultivars reliable and create presumptions for their commercial success.

Jõgeva PBI is the only institution dealing with plant breeding in Estonia. The varieties bred out at the Jõgeva PBI are the intellectual properties belonging to Estonian nation and is the guaranty of economic independence in agriculture.

2. Estonian Research Institute of Agriculture

The Estonian Research Institute of Agriculture (ERIA) was founded on July 1994 as the result of reorganisation the former Estonian Research Institute of Agriculture and Land Improvement (ERIALI) established in 1956.

Subjects of scientific and development activities on potato of this Institute are:
- Improvement of growing and nutritional conditions of potato and the increase of growing profitability by additional agro-technical methods.
- Studying the agro-technical methods increasing marketability of potato, nutritive value and the duration of consumption period.

- Improvement of integrated plant protection system on growing potato, considering the requirements of environment protection, a good agricultural practise and economic aspects.
- Modelling of production process and yield of agro-ecosystems.
- Potato storing and duration of consumption period.
- Effect of agrotechnical methods to the phytosanitary situation of potato fields.
- Finding out technologies and machines for potato production suitable in the conditions of Estonia and the economic efficiency of their use.

3. Estonian University of Life Sciences

Estonian University of Life Sciences is the only university in Estonia whose priorities in academic and research activities provide the sustainable development of natural resources necessary for the existence of Man as well as the preservation of heritage and habitat.

The spectrum of disciplines taught at the agronomy is wide. The students of agronomy must be involved in some kind of research work and this gives them an opportunity to defend MSc thesis. Main items of research activities on potato:

- The physiological ageing of seed potatoes and it influence on formation of tuber yield and quality.
- The timing of the potato late blight control achieved by use of Hardi Metpol and in the internet based decision support system (DSS) NegFry.

4. Research Centre EVIKA Tallinn University of Technology

The foundation year of Plant Biotechnological Research Centre EVIKA is 1966. This year the first attempts were made to study the virus eradication of potato and carnation plants in Saku. At that time only few successful potato meristem cultivation trials were reported in the world.

At the present the main research areas on potato are:

- The studies of potato meristem characteristics and somaclonal variation of meristem clones. The main task of research work is to identify the relations between meristem characteristics, methods of eradication technology and somaclonal variation. The identification of above-mentioned relations enables to direct the results scientifically, increase the disease resistance and improve the economically important properties of varieties as well as shorten the breeding process.
- The studies of resistance problems of potato varieties and breeding lines. The aim is to study the resistance of potato meristem plants and clones to various races of pathogen *Phytophthora infestans in vitro* and in field conditions. The use of meristem plants in identification of *Ph. infestans* races is the unique approach in research.
- The research of productivity of potato meristem plants are focused on the discovery the correlations between the yield and tuber's size of potato meristemplants grown in field depending on *in vitro* and *in vivo* plant propagation method, leaf area index, intensity of photosynthesis and duration of vegetation period.
- The preservation of potato genetic resources. The goal is to study the correlation between the conditions of long-term preservation of plants and morphological characteristics and economical traits of varieties and cultivars.

Conclusion

Agricultural reform, the market situation and structural changes in enterprises have influenced the development of potato production. Potatoes are mainly grown in farms and households and also to a lesser extent in agricultural holdings - 5% of the entire area under potatoes. The area under potatoes has decreased in the recent years. At present the food potato demand is fully satisfied - 120,000-140,000 t (one person consumes ca 100 kg of potatoes annually on average). Only certified seeds of early and mid-early varieties and varieties of industrial potatoes need to be purchased from abroad for further propagation.

In the future our main strategic goals in potato production are:
- to fully cover the population's needs by domestic potatoes;
- to improve efficiency and competitiveness of potato production and bring into compliance with EU requirements;
- to guarantee stability of domestic market price level;
- to maximize the agricultural sectors contribution to national economic and social well being on sustainable basis.

The Estonian Research and Development Strategy "Knowledge-based Estonia" 2002-2006 was approved by the Parliament in December 2001. On the basis of this strategy the government approved the national programme "Applied research and development in the field of agriculture 2004-2008". In this programme the priorities are given to the control of plant and animal diseases, reduction of usage of pesticides and preservation of genetic resources and biological diversity. Driven from abovementioned priorities the potato research is currently focused on:
- breeding and selection of new high-quality disease-resistant varieties;
- improvement of selection and initial seed production methods;
- development of environmentally friendly and resource-sustainable production and storage technologies.

www references

1. http://www.agri.ee/Overviews/Agriculture and Rural Development overview 2003, 18.04.2006
2. http://plant.agri.ee/Seemnete sertifitseerimise andmebaas SERTI2/Põldtunnustamine 2001-2005, 2005, 17.04.2006 (The seed sertification database SERTI2/Field approbation 2001-2005)
3. http://www.stat.ee/Statistical Database/Agricultural Census 2001/Crop Production/Use of agricultural land/ACT201/ Use of agricultural land in agricultural holdings by legal form of holder and use of land, 2001, 18.04.2006
4. http://www.stat.ee/Statistical Database/Economy/Agriculture/Crop Production/AGS113 Growing potatoes by legal form of holder and size class of shown area, 2005, 18.04.2006
5. http://www.stat.ee/Statistical Database/Economy/Agriculture/Economic accounts of agriculture/AG63Value added and entrepreneurial income in agriculture in basic prices of the corresponding year, 2004, 18.04.2006
6. http://www.stat.ee/Statistical Database/Economy/Agriculture/General Data/AGS402 Structure of holdings by legal form of holder, 2003, 18.04.2006
7. http://www.stat.ee/Statistical Database/Economy/Agriculture/Labour force/AGS201 Labour force by country and kind of labour forse, 2004, 18.04.2006

Curriculum vitae – Katrin Kotkas

Author K. Kotkas - born at a farm workers family - studied at the Estonian Agricultural Academy from which she received a MSc in 1994 based upon her work on influence of medium content on regeneration of potato plants from microcuttings *in vitro*. In 2003 she received DSc degree from Estonian Agricultural University. The thesis explained the influence of culture medium composition and preservation conditions on *in vitro* preservation of potato varieties by means of meristemplants and on subsequent performance in the field. For the preservation of potato varieties as meristemplants the preservation medium and conditions that did not have negative influence on the quality of *in vitro* plants and on the productivity in the field, were worked out. For the first time the information about the productivity of the meristemplants from the field depending on the content of medium and growing conditions *in vitro*, was obtained.

K. Kotkas has been working in Plant Biotechnical Research Center EVIKA for almost 25 years on several professional positions. She has improved her professional knowledge in courses of preservation of plant genetic resources in Hungary (1997), in Italy (1998) and of biotechnology in Japan (1999). At present she is a leader of 5 applied and innovative research projects. She has published 87 scientific and scientific-popular papers and is co-author on 2 SU patents. She is active member of many agricultural scientific and development councils in Estonia.

Company profile – Research Centre EVIKA Tallinn University of Technology

Plant biotechnology is a priority research field in the whole world. The plant biotechnology research in Estonia was started in 1966. The main aim of EVIKA research projects is to study the eradication of plants from virus diseases, increasing the disease resistance and yield capacity of plant material; creation of highly effective eradication, propagation and production systems for various plant species and cultivars. Meristem clones with improved characteristics, disease-free, true-to-type plant material for seed production or nurseries are the result of successful research work. The scientific studies involve also the preservation of plant genetic resources and biological diversity using tissue culture methods.

The research of eradication and propagation of basic seed potato material for Estonian varieties is done only in EVIKA. EVIKA participates in several international projects and has a close collaboration with many research institutions and production firms, especially in Nordic and West-European countries.

General information on the potato production, trade and perspectives of potato market development in Russia

V.D. Molyanov
Solana-Agro-Service OOO, 45A A. Tolstogo St., office 201, Samara 443099, The Russian Federation

Abstract

Potato is a highly profitable crop. Russia is forming a market for this product at the moment. A network for production, sale, packing, washing and packing, and processing is dynamically developing in Russia. The most advanced companies of this industry provide high quality products and work with facilities in accordance with world standards requirements. At the same time there are problems of know how transfer to the farms a shortage of qualified personnel, machinery and high quality seeding material. Resolutions of these problems will allow the Russian potato sector to achieve sustainable and high yields.

Keywords: variety, technology, Solana, farm types

Introduction

Potato has been grown in Russia for a long time. Since the time of Peter the Great this crop is known in Russia first as an ornamental plant and then as an irreplaceable product in Russian people's life. People often say in Russia: Potato is a substitute for bread. And it is really true, though potato products consumption is decreasing from year to year, consumption per capita still exceeds 100 kg per year. The Potato is being replaced on our menu by other products such as fruits and vegetables. They are becoming more accessible due to the development of new technologies, varieties, logistics and transportation systems. The world is moving to the stage when people prefer to consume "healthy" food. However potato producers are not discouraged: the demand for these products is stable, supermarket networks are expending, processing systems are developing. Problems with overproduction were not noticed during the last years.

Production situation

General condition and production volumes

Russia is a big country, agriculture and potato production is practiced practically in all regions including Sakhalin and Kamchatka. Basic areas of this crop production are concentrated in moderate climate area of the Russian Federation. Here black soils, Chestnut soils and grey woods soils prevail where during vegetation period precipitation amounts to 200 to 350 mm. Major regions of potato production are Moscow, Tatarstan, Leningrad, Chuvashia and the Sverdlovsk regions. The biggest concern of potato producers is the lack of water, more than half of the areas are located in regions with limited rainfall. Moisture is a limiting factor for potato yield. New watering and irrigation systems are required. During last years the area for potato production almost did not change and amounts to over 3 mio. hectares, (see Table 1).

Table 1. Potato production in the Russian Federation, 2004-2005 (Source: Federal State Statistic Service).

Total yield, (mio. T)		Average yield, (dt/ha)		Area planted with potatoes (mio ha)	
2004	**2005**	**2004**	**2005**	**2004**	**2005**
35.9	37.5	115	122	3,150	3,088

In 2005 the share of large scale on farm potato production of was only 8.4%. The rest of the potato was grown on the personal plots of rural people. For the last years there is a trend of reduction of such plots. According to different experts and specialists this development will continue for several more years. Often plots or gardens are located far away from residential areas or plots owners decide to use land for rest or for lawns *etc.* due to the family budget increase. The economic inefficiency and the low yields on a private plots result from poor crop management and low quality seed. According to Federal State Statistic Service average yield on the small plots was 11.9 t/ha in comparison with larger farms averaging 15.6 t/ha.

At the same time there is an increasing trend of the interest of the bigger farms and investors to increase the areas planted with potatoes. This is related to the stable prices and the high demands for fresh and processed potato. Accordance to Samara-Solana ZAO data, the rate of return of field potato production is 30- 45%. There is a trend towards reduction of the profitability of field production due to the increased costs (diesel fuel, irrigation, inputs). Additional profit can be added by investment in storage facilities. After proper storage the annual profitability of potatoes can be as high as 70 - 80%. During the last years new large-scale producers of potatoes appeared, *e.g.* Svekla 000 with over 800 ha of potatoes, Bahetle Agro 000 with 500 ha (Tatarstan), Gorby (KHF) with over 2500 ha (Belgorod), Krimm 000 with over 1,000 ha (Tumen), Dmitrovskiye ovoschi with 400 ha (Moscow), Luch 000 with 500 ha (Krasnodar), Chernosemje Agro with 300 ha (Voronezh). Quite often they are even not presented in Top 100 rating of potato producers (see Krestyanskiye Vedomosty newspaper, volume No. 9, February 2004). I would like to name proudly farms included in the first 10 of the rating, such as Number 1 - KRIMM 000 with average yields for 3 years of 30.1 t/ha, Number 2 Vega 000 with 40.4 t/ha and Number 6 - Lunacharsk SZAO with 23.1 t/ha. The companies use Solana varieties in their production.

Trends of the industry development

Rapid areas increase, of more than 10% per year is impossible because of the lack of sufficient amounts of suitable machinery, seed material and potato storages. Therefore, due to these factors the statisticc data show that the average potato yields are not changing significantly from year to year and remain at a low level of 10.0 -12.0 t/ha. Table 2 shows the reasons for the low potato yields.

With production areas of 3 million ha, as we indicated earlier, the annual requirement of seed potatoes should be over 6 million tons, if we assume a seed rate of 2.5 mt/ha. Market of seed material in general for Russia including seeding material brought from abroad is limited to 200 - 220 thousands tons with a share of imported seeds of less than 10,000 tons. This is

Table 2. Reasons for low potato yields.

1	Field location
	a. Short vegetation period
	b. Heavy soils
	c. Lack of moisture
	d. Hot climate
2	Low quality seeding material
3	Varieties are not suitable for production with agricultural machinery
4	Insufficient amount and/or low quality of machinery for production, treatment harvesting and storage
5	Insufficient level of knowledge
6	Lack of storages and poor conditions for seed material storage

absolutely insufficient. Some farms reproduce their own seeds and many farms simply use low quality degraded seed material for planting. At the moment the protection of plant breeders rights seems to improve with Rosselkhozhadzor control (Russian State Agricultural Supervision Service), that is why the demand for the certified seeds is expected to grow.

Practically all large scale farms build modern potato storages, for example in the last two years the following new storages were built in Samara oblast: Agronix OOO – 3,000 mt, Samara-Solana ZAO - 5,000 mt; in Voronezh oblast: Chernosemye Agro – 4,000 mt tons; in Tatarstan - Bahetle Agro OOO – 12,000 mt and in Tumen Oblast - Krimm - 20,000 mt. Many companies plan to build new storage facilities (In Samara oblast Stepana Razina farm – 10,000 mt, in Moscow oblast Zhilevskoye OOO for 5,000 mt). Investments in this sector can be estimated to 300 - 500 EURO per square meter. Additional expenses are required for insulation against high and low temperatures. There is a trend for the producers to trade potatoes all year round. Also there is an increasing trend to pre-pack and pack potatoes. For these purposes farms install new grading and packing equipment in the storage facilities. The most well known companies in this sector are: Dmitrovskiye ovoschi, Dmitrovskoye selo, Agrotrade, Tulskaya Niva.

Further problems arise with the absence of sufficient numbers of modern agricultural machines for planting, treatment and in particular for harvesting and storage of potato. According to my estimation, the annual requirement for potato harvesters such as the Grimme DR 1500 type is more than 100 units per year. Even this amount would be able to harvest not more than 150 thousands ha per year. Many farms still use old technologies and machines for potato production. High profitability of production allows large scale farms with minimum potato areas of 100 ha to invest their funds in buying new machinery. Reports of foreign and domestic companies show an increase of the sales volumes.

Traditional technology using domestic machinery is designed for 70 cm inter-row spacing. In the summer time it can get very hot in the central part of Russia. The potatoes may get overheated in the small ridges, what may result in a low number of tubers per plant. If weather conditions are favorable and yield increases to over 30 mt/ha in small inter-row spacing lots of tuber turn green, what will cause quality problems. Many farms also use 75 cm spacing technology using mainly European machines or machines assembled in Russia under license of the foreign companies. For instance, quite interesting work is done by Eurotechnicka ZAO, Samara, a factory producing Grimme and Amazone machines for potato production as well as Kolnag factory, Kolomna which produces Dutch machines of the same type. These types of machines are more adapted to the needs of local producers plus they provide service for the machines. In Southern regions such as Krasnodar, Rostov and Stavropol potatoes are grown with 90 cm spacing.

As our company's experience, which is based on training and seminars with farms specialists and agronomists and on consultancies on site, shows it is possible to increase the yield only by using new knowledge. These services are in high demand at farms. Due to transition period in the economic reform the influx of qualified personnel to this industry was very low. This is also valid for the large scale and highly efficient farms.

An excellent example is the success of the Samara oblast being the leader in the introduction of advanced and modern technologies in potato production. For instance the average yield of potatoes in the last year was over 22.1 t/ha, that is almost twice as high as the national average and is the best yield in Russia. The best farms of year 2005 such as Vega OOO with 49 t/ha yield, Samara-Solana ZAO with 41.0 t/ha get stable results every year. The Povolzhye area is considered to have a dry and arid climate and history knows examples of complete failure of the crops in this area. Taking into account the dry climate conditions potato is grown under irrigation here. Production intensification, economical calculations and new knowledge are the components of modern and flexible technology for potato production, which is adapted to different soil conditions. A modern unique approach to the introduction of new technologies consists of 4 components:
1. technical - high productive and reliable machinery use;
2. biological - using high quality seed material of varieties adapted to mechanical treatment;
3. agrochemical - efficient plants protection and fertilizer use;
4. consulting - new knowledge and know-how use.

Today, the first component is resolved through the production of modern agricultural machines for potato production by Eurotechnicka ZAO, the second through the production of high quality seed material of regionally adapted varieties in Samara-Solana ZAO, the third component - through agrochemicals produced by Syngenta, BASF, Bayer and others and the forth component through training of the specialists on Samara oblast farms.

Variety availability and seed production

Many scientists say that none of developed technologies, even the most efficient ones can be automatically transferred to another region. It has to be adapted to the soil and climate conditions of the area, its components have to be tested and studied in the actual production. This statement also true for varieties. There is a large group of so called flexible varieties suitable for growing in different regions. This group includes such varieties as Nevsky, Lugovskoy, Rosara, Secura, Sante and others. There are potato varieties which are, due to day length, vegetation time and sum of temperatures above zero, more risky to grow, what might result in yield loss. The State Register of varieties in the Russian Federation includes more than 180 varieties, 70% of these varieties were bred in Russia. However farms grow or use only about 30 varieties.

According to data, see Table 3, based on Federal Seed Inspections service data, the most important variety is Nevsky, which is planted on more than 40% of the potato area. Among the foreign varieties Romano (Agrico) and Rosara (Solana) are the most commonly used ones.

Because of the rapid development of the potato processing industry the grower's attention is directed to varieties suitable for these purposes (Saturna, Lady Rosetta, Hermes, Velox, Panda, Innovator and others), the market is developing in this direction. However, there is not enough supply from the companies of this type of potato. Producers relate the reason of low supply with extra costs for production, for example with extra costs for special fertilizer or with the prolongation of the vegetation period of the chips varieties which is risky for harvesting or with expensive conditions for storage. Thus the processing companies have to import additional volumes of potatoes.

Potato import volumes are focused on the import of seed potato, potato for processing, table potato and early potato. Accordance to Federal State Statistics Service data import volume amounts to 40-50 thousands tons of potato per year.

Due to infection pressure not all areas are suitable for potato production. Another problem related with quality is that many farms do not observe the basic requirements of seed potato production such as:

- the exclusion of quarantine diseases (potato wart disease, nematodes);
- crop rotation (minimum 4 years);
- high quality seed use;
- seed treatment/warming/pre sprouting before seeding;
- separation of seed crop areas from consumption areas (minimum 2 km);
- use of natural shelter belts (water storage basins, forests);
- use of tramlines for the application of agrochemicals and elimination of mechanical treatment after emergence to avoid mechanical virus diseases transfer
- application of insecticides against vectors every 7-10 days;

Table 3. Varieties, leading in seed production volumes, year 2004 (Anisimov, 2004).

Rank	Variety	Quantity, t
1	Nevsky	89,141
2	Udacha	15,890
3	Romano	14,113
4	Lugovskoy	12,204
5	Elizaveta	11,577
6	Rosara	11,163
7	Peterburgsky	4,022
8	Sante	3,193
9	Zhukovsky ranny	3,169
10	Karatop	2,935

- removal of diseased/infected plants;
- weed/disease/pest control;
- timely haulm killing;
- careful harvest to avoid damages;
- intensive wound healing period;
- temperature during storage should be kept at the level of 2-4 Celsius degrees to avoid physiological aging of the tubers.

The following companies are known in the potato seed industry with seed volumes of over 1000 tons: Agrico-Eurasia, Doca Genniye technologii, Samara-Solana, Solana-Agro-Service, Norika-Slavia, Meristemniye kultury, Slava Kartofelu, *etc.* A system for potato seed certification including laboratory tests for the presence of viruses was introduced. In Samara oblast all seed lots that are used for planting in large scale on farm production are subject to virus control.

Trade and potato market in Russia

The purpose of production

The purpose of potato production is directly connected with ware potato markets. We can distinguish three basic areas of production:

1. the southern regions -early maturing ware potatoes for fresh consumption;
2. central part of Russia and Siberia - table potato for storage;
3. Moscow oblast and adjacent areas - table potato and for processing.

At the moment basically all regions due to low level of processing grow table potatoes. Less than 10% of the total volume of production is for processing. There is almost no production and consumption of salad potato (cooking type A) in Russia. Consumers prefer varieties with red a red skin, creamy to light yellow flesh, cooking type B and with an excellent taste. For these red varieties 5-10% higher prices are often paid in comparison to yellow skinned varieties. The sales price is often based on the visual appearance and not on other criteria such as cooking quality or taste. In the stores varieties are neither separated between different cooking types nor for their suitability for *e.g.* mashed or fried potatoes or French fries. Also the names of the varieties are not printed on the labels. In big supermarkets such as Metro, Perekrestok, Ramstor and others, brand names appear with names of the producer or packer - buyers relate quality of the product, packing and safety to this name.

Large volumes of food potato are sold and bought in September - October in Russia. It is connected with the traditional purchase of food by the people for the whole winter and with the purchase by big companies for their own stocks. Due to lack of the storage facilities adapted to winter conditions, quality decreases often in the winter time. For example many suppliers refuse to load potatoes at temperatures of -20°C outside. This results in price increases. That is why big suppliers for supermarkets networks have to store big volumes of potatoes in their own facilities.

Major consumers of potatoes for processing, mostly for potato crisps are Pepsico in Kashira, Bridgetown in Vladimir, Kunzevo in Vologda and Russky product in Moscow. At the moment these companies are negotiating the expansion of their production facilities. The production of potatoes for French fries and for quick frozen products is at the development stage now. Such companies as McCain, Sibirsky Bereg, Gorby, KRIMM stated that they are planning to process these types of potato.

Prices and potato market

The early potato market was almost not developed until the beginning of the year 2000. The reason for this was the low demand and the low purchasing power of the population. In trading networks similar to Europe the first fresh potato appear in the beginning of the year, of course these products are imported from Egypt or other warm countries. The first domestic potatoes appear at the end of May or beginning of June. Prices are high at that moment and there is no stable demand as potatoes from the previous year are still in good condition and due to late spring in Northern regions and it is offered to the consumers at a low price. The highest demand for potatoes is in July - producer's prices vary then from 6 to 12 rubles per kg. Then by fall, for example in last September during potato harvesting season the price was 3.5 - 4.5 rubles per kg. These prices were at the same level in fall 2005. Preliminary data show high yields and area increase in comparison with the previous year. Indeed gross yield in the farms was 37.5 mio tons, that exceeded results of 2004. However demand exceeded offer and already in the end of November potato prices went up sharply and by March 2006 they reached a level of 10-11 rubles per kg. To my estimation the trend for high prices will continue in the near future as already now there are preliminary contracts for the yield 2006. Traditionally prices are determined by the demand from central regions of Russia: Moscow, St. Petersburg, Nizhny Novgorod. In case of overproduction of potatoes in other regions, it may happen that with increasing distance from Moscow the prices will drop on the local market. For comparison in the beginning of October 2005, the price for potatoes in Moscow was 4.0 - 4.50 rub/kg, in Samara it was 3.50- 3.80 rub/kg, in Ekaterinburg 3.0-3.20, in Kurgan and Novosibirsk it varied from 2.0 - 2.5 rub/kg. Up to now potatoes are traded within the central regions. According to estimate of specialists a minimum of 80% of the potatoes supplied to regional supermarket networks located from Moscow up to 1,000 - 1,500 km away come from Moscow. That means that the potatoes are returning back to the regions. The transport of potatoes to Moscow is cheaper than transportation into the other direction as the trucks are returning back empty. From Moscow more potatoes are supplied to regions than in the other direction back. Therefore some of the trucks are returning empty. The difference between cost of potato in supermarkets and from the supplier sometimes is more than 100%.

The most important consumers of potato produced on the small plots are the producers themselves. They sell only surpluses not used for their own needs. Within the National Project for Agricultural Development the state represented by executive authorities is trying to mobilize the tremendous potential of the smallholders for consumers market supply.

Potato produced by large scale producers and farms is supplied to:

- stores, cafes and restaurants;
- whole sale markets for further resale;
- big logistic centers for packing, washing and pre-packing for supply to supermarkets networks;
- public organizations through tenders (schools, kinder gardens, rest homes);
- big industrial companies which purchase potatoes for their employees;

- processing companies;
- state organizations like army, prisons *etc.*

The seed potato market is more chaotic than the ware potato market. The slower potato varieties degrade the longer and more it is used on the farms. For example the variety Nevsky is commonly used by farms as repeated reproduction. Many good foreign varieties are not protected by patents due to the late registration (more than 4 years) from the moment of its first sale in the world. Therefore the client who purchased seeds one year will not necessarily be the client next year. The highest demand of the consumers is directed to early maturing table varieties. The vegetation period together with the lack of moisture are the limiting factors for the production as many regions are located in climate zones where the vegetation period is limited to 100-120 days. Prices for seed material vary; they depend on the volumes, dates of the sales, quality of seeding material and reproduction level. In the fall 2005 they were 6-8 rubles per kg for Grade A potato and 10-15 rubles per kg for Grade E potato. Before the seeding season of 2006 seeding material was in high demand and supply was limited, therefore the prices went up to 15 rubles per kg for class A potato and to 20 rubles per kg for class E potato. Import of the seeding material was limited in 2005 spring due to quarantine bans and in spring 2006 it was reduced because of increased transportation cost. Prices for imported seed of class E varied from 24 to 28 rubles per kg after customs clearing depending on the variety.

Conclusion

With respect to the above facts the Russian market seems to be quite interesting both for domestic producers and for the foreign investors. Because all production components such as seed, agricultural machines, agro chemicals and the latest production know-how are in high demand there is a unique opportunity to offer your services and products in this most interesting sector - large scale on farm potato production. Dynamic developments are expected also in the construction of storage facilities, packing, pre-packing and processing. Also in Russia according to the consumer's estimates, new very early maturing variety with red skin with high resistance to diseases would be very successful.

Acknowledgements

I would like to express my gratitude to my teacher, author and to the person who introduced new technologies of potato production in the Samara oblast - to Dr. G.I. Böhmig, to the manager of Eurotechnicka ZAO company Mrs. L.V. Orlova and to the managing shareholder of SaKa-Ragis Dr. Kartz von Kameke, for their help in the organization of the joint project for seed potato production in Russia.

References

B.V. Anisimov (2004), Kartofel i ovoschi [potato and vegetables] magazine, volume 4, p. 25

Federal State Statistic Service, Russian State Statistic Service data, (2005), http://www.mcx.ru

Top 100 rating of the potato producers, Kerstyanskiye vedomosty newspaper, vol. 9, Febryary (2004).

Curriculum vitae – Vladimir Dmitrievich Molyanov

Vladimir Dmitrievich Molyanov was born in 1970 in Pokhvistnevo region, Kuybyshev oblast of the Russian Federataion. He graduated from Samara Agricultural Institute in 1992 specializing in agronomy. Immediately after graduation he worked as agronomy expert for modern technologies of potato production in the Samara Potato Project developed by Management and Production Systems (MPS) Ltd., Samara. As a consultant he introduced new agricultural practices and new agricultural machinery in different farms of the Russian federation. During 1995-1999 he had a postgraduate scholarship in the All-Russian Scientific Institute for potato farming, Moscow, specializing in plants production. In 1999 he received a PhD degree in agriculture for his work on influence of different rates of mineral fertilizer on yield and quality of potato grown under irrigation in the Povolzhye region. Since 1995 he has been working as a project manager to develop a company for seed potato production together with the German company "Solana Agrar Producte GmbH KO KG". In 1997 V.D. Molyanov became technical director and in 1999 he has been appointed as director general of the newly established joint venture "Samara-Solana" ZAO. This enterprise is one of the largest potato seed producing farms in Russia. The annual potato production exceeds 5 thousands tons. During last three years the average potato yield exceeded 40 t/ha. In 2003 Samara-Solana ZAO was nominated as the company of the year in the category "High efficiency in agricultural production". In 2002 V.D. Molyanov also became director of "Solana-Agro-Service 000", a company that represents the Solana Agrar-Produkte and Saka-Ragis potato varieties in Russia. He is author of numerous scientific articles and presentations on conferences.

Company profile – Solana-Agro-Service

Solana-Agro-Service 000 is the general representative of the German company "Solana Agrar Producte GmbH KO & KG", Hamburg, Germany in the Russian Federation. Solana is the exporter of potato varieties bred by Saka-Ragis and offers high quality potato seeds. The varieties Rosara, Secura, Arosa, Panda, Velox were included into the State Register of The Russian Federation and are also listed in different regions. "Solana-Agro-Service" 000 produces seed material based on reproduction and license contracts on an area exceeding 1000 hectares under careful supervision of the company's specialists.

The Solana is a very important company working worldwide on seed potato marketing. Solana markets the varieties bred by SaKa-Ragis Pflanzenzucht Gbr, Hamburg and DEN HARTIGH BV, Emmeloord. SaKa-Ragis ensures potato quality supplied from Germany and DEN HARTIGH supplies high qualitative seed material from Holland. The integration of the German breeder SaKa-Ragis and the Dutch breeders DEN HARTIGH into one network has made available a big range of good varieties for the customers. This seed material is exported from Hamburg to all over the world. Samara-Agro-Service" 000 is also a representative of the Dutch company Nickerson-Zwaan in Russia. Providing complete information on these varieties our company supports the customers. Further SOLAGRO is delivering consultancies and organizing seminars on potato production during the whole vegetation period.

Changes on the German potato market

Christoph Hambloch
ZMP Zentrale Markt- und Preisberichtstelle für Erzeugnisse der Land-, Forst- und Ernährungswirtschaft GmbH, Rochusstraße 2, 53123 Bonn, Germany

Abstract

A detailed overview upon the German potato market is given. Export of raw and processed potatoes is growing, whereas absolute consumption declines, effecting the whole potato chain. Structural changes on the farm level lead to a concentration of the potato production. With respect to limited storage facilities, a huge amount of stored table potatoes are imported competing with early potatoes from the Mediterranean. A diversification of the potato processing industry also guides to specific market situations including import and export levels.

Keywords: potato market, export, import

Within the EU-25, Germany is one of the largest potato producing countries. Of last years nearly 2 million ha of potatoes in the European Union Germany had a share of 14% (or 277,000 ha). As yields are far above those in other countries, Germanys share of the total production (58.6 Mio. t) is nearly 20%. Although only a few of those are destined for the early potato market a harvest of half a million tons represents nearly 18% of the production within the EU-25. Potatoes for the production of starch have a big share of the total German crop. Not everywhere in the European Union starch potatoes are grown, so that Germanys 2.5 million t last year made 36% of the total EU-starch potato production. Because of its large population Germany is one of the major destinations for imports from countries on the Mediterranean. Especially early potatoes from Egypt find their way to Germany very voluminous. But also deliveries from Israel are getting larger and larger every year. Although it is the image of the Germans they do not eat most potatoes compare to people in other countries. Especially in the north of Europe but also in Spain, in the Baltic countries, Malta and Poland the per capita consumption of potatoes is higher. Of the 68.5 kg per capita in 2005 quite a big share were potatoes in processed form like French fries, crisps, dumplings or mashed potatoes.

The use balance for potatoes shows different trends in Germany over the past years:

- Exports of fresh potatoes and potato products are rising every year. Lately the additional opportunities in the East European countries but also in countries like Russia, Ukraine and so on pushed the figures more upwards. In 2005/06 the German trade had the chance to fill the holes that where left on the eastern potato markets as the harvest there was quite small. In contrast to the potato production in these countries German traders are able to deliver big quantities in consistent quality. As retailers in East European countries are mostly based in the west, they have similar requirements like in their home land which can not be fulfilled with the potato production of the local market in certain times of the year. This opens the door to deliveries from Germany. Compared to France, Belgium or the Netherlands exports of potatoes are still small.

- The consumption of potatoes is declining every year. Since this decade the declining of fresh potato consumption is no longer compensated by an increased consumption of products. Therefore, the total use of potatoes for human consumption declines.
- The potato harvest for starch production is depending on EU-regulations, so that there is not a market development in the traditional way.
- The biggest change in the recent years is the concentration of potato production on less farmers. From the year 1993, when nearly 160,000 farmers produced potatoes the number of farms has halved. As area didn't decline in the same way, each farm now produces on more hectares. Structural changes in Germany are very different, depending on the region. In the most important "Land" Lower Saxony, within nearly 10 years the area per farm has risen by about 60%. One can find similar developments in the west of Germany. In the east, where farm size since the "DDR" has been much larger than in the west, the changes are not that strong. The slowest development is in Bavaria, where we still have a lot of very small scaled farms.
- With the concentration of potato area on specialised farms and most suitable soils the certainty of yield should rise. This is by farm not true. Depending on the time of planting, temperature, and the water supplied during the season, yields can change a lot. Not in every place farmers irrigatr. This production tool is wide spread for the early potato production and in the "Heideregion" of Lower Saxony, where a lot of table potatoes are grown. In the east, in Bavaria and in the south-west irrigation is more rare. Certainty of yield could rise because of a warmer climate, but reality is more complex. The global warming brings more extreme atmospheric conditions as drought, floods or mild winters with early sprouting or pests in the following year. One must accept that extreme fluctuations in potato harvests will be the future.
- Talking about old crop potatoes, traditionally the most important origin for the German market were the Netherlands. For the fresh market the meaning has been declining until the nineties.
- During the last years the deliveries for packing from cold stores got more and more important. As in Germany there are only few cold stores, most of the potatoes sold from the second half of April until the beginning of June have to be imported. The Netherlands have their share of it, but more and more imports are from France. This year the total amount of deliveries could well be about 100,000 t.

There are several reasons for the increase of potatoes from cold stores from West European neighbouring countries. German farmers didn't invest much money in cold stores so far. As a consequence, there is only a small availability from April onwards. In France, there have been strong changes in the areas of production. Bintje used to be the main variety also for the fresh market until the 80th. Since than diversification took place, so that nowadays many varieties which are also suitable for the German market are produced. Bintje has been grown in the north of France, but table potatoes have now a home in regions like Beauce or the Champagne. There best soils which have not been used for potato production before have become available, and the very good qualities produced there have have been stored under optimal conditions from the very start. This was the basis for a strong expansion of exports to other countries. Potatoes from French cold stores do not look much different than early potatoes from countries on the Mediterranean. They also have a very bright skin without any damages so that because of the much lower price they have a competitive edge. In Germany two sides of the market have been surprised by the latest developments: Those traders that offer imported early potatoes find it more and more difficult to place their products during the early stage of marketing. Farmers now also have their problems from March onwards to compete with the beautiful looking French potatoes. There have been efforts to increase quality programs with cold stored potatoes in some places of Germany, but the amount of potatoes coming from there still is not sufficient to meet market requirements. Most probably this will change in the future.

The German potato market is one of the main destinations for early potatoes from countries on the Mediterranean. Up to 40% of deliveries from third countries arrive here. The main source is Egypt with about 100,000 t. But there is one major problem with these potatoes: Because of the fear of being banned from the market due to infections with quarantine diseases and because of the duty free quota which is closed at the 1st of April all supplies arrive here in March. The "earlies" are put into cold stores in North Italian, North Germany, Belgium and in the Netherlands in March. At this early time of the year the demand for early potatoes is only very small. Huge quantities have so be sold later from May to June. This doesn't go along without changes in quality. Here deliveries from Israel are in advantage. Early potatoes from there can be delivered just in time which usually has the effect of a much brighter skin. As the German retailers more and more want best looking tubers, the market for Israeli potatoes has been growing recently. Earlies from Morocco are produced on very small scaled farms with the result of strong alternating qualities. The preparation in 15 kg-bags is another issue of problem. In the opposite direction - exports of table potatoes to third countries - Germany plays an inferior role. Even during the marketing year 2005/06 there was only little export to Russia or Ukraine, where the Belgium and Dutch industries where able to place larger quantities. On the contrary, destinations like Italy, Poland or Southeast of Europe seem to be picking up more table potatoes from Germany each year.

The development of the potato consumption in Germany shows a very sad pattern. Changing lifestyles, declining number of big families, higher employment rate of women and the old fashion image of potatoes have led to a decline of the per capita consumption of fresh potatoes to only 33 kg per year. The same level or even a little bit more potatoes are consumed now in form of different products like French fries, dumplings, crisps or mashed potatoes. During the nineties the extra use of potatoes for processing compensated parts of the decline of table potatoes sales. Since 2001/02 potato processors observe new trends of a healthy and balanced diet. Obesity among children and a general discussion about certain ingredients of potato products led to a smaller consumption of potato food.

For the fresh market the analysis of GfK (Purchase of german households by GfK (Association for Consumer Research), Nürnberg) shows a more detailed and clearer picture of the latest developments. Within 10 years the potato purchase of private households has dropped by nearly 23% to only about 1,4 million t for the whole of Germany. There are no signs on the market that this development will be stopped within the next years. Of course there are consequences for the market. The packing capacities have to be reduced which usually is a progress with strong competition on the market.

Of course the competition is also getting harder along with the concentration of food retailers. But it is not only the concentration but also the mentality of German consumers to buy as cheap as possible. This boosted the development of discounters like Aldi and Lidl in the German food retailing. Nowadays more than 50% of all fresh potatoes are bought in those kind of outlets. As they have a very lean structure, the market has become very transparent. The diversity of the potato offered has been afflicted. Consumers don't find a wide offer of different packing sizes, of varieties or cooking types.

The diversity among potato products on the contrary has been growing. Of course French fries are still one of the major products, but other deep frozen potato food as well as different chilled preparations have become more popular recently. The German processing industry is by far not able to produce all the deep frozen products which are needed on the German market. Many French fries come from the Netherlands or from Belgium and they make up to 50% of the supply on our market, while the situation for other products is different. Because of the logistics, chilled food from Germany has an advantage. Another sector of the industry is producing dehydrated food stuff, *e.g.* for the production of dumplings, mashed potatoes, soups and snacks. This industry has been very successful in exporting its products, but since 2004/05 there seems to be a little decline in sales and export.

References

Produciton and foreign trade figures from Statistisches Bundesamt, Wiesbaden, 2005 and 2006 and national official statistics from countries mentioned.

Hambloch, C., Menth, H., and Stelzer, M. (2005). ZMP Marktbilanz Kartoffeln 2005.

Curriculum vitae – Christoph Hambloch

Although growing up on a farm on the fertile loess soils in the Rhineland, Christoph Hambloch got into contact with the potato market not earlier than at the beginning of the nineties. Before that he enjoyed the university education in agricultural science, branch plant production, at the Friedrich Wilhelm University of Bonn. At the same time Christoph Hambloch also managed the farm of his father. In January 1992 his career at the ZMP started. First he was responsible for the development of a price reporting system for the potato brand "AckerGold". Practical experiences with the market grew by his marriage into a potato trading family in Cologne. Right from the beginning of his work for ZMP the potato market was the focus. Beside the market reports lectures, advisory boards and associations - both national and international - are since then on the regular schedule. Later, in 1996, he took over the management of the department for arable products in the company. At the same time he became secretary of the union of Early Potato Growers Associations. Together with this group Christoph Hambloch enlarged his contacts to and knowledge of the European potato industry by organising regular excursions to different countries, meeting the potato industry there. Another engagement is for the potato futures market. Together with colleagues he published a handbook for beginners and professionals in using the future market for hedging.

Company profile – ZMP

ZMP was founded in 1950 as an agricultural service institution and focused its activities in 1972 on agricultural market information. It acts on behalf of the Absatzfonds according to the Absatzfondsgesetz (Agriculture Marketing Fund Law). In 1990 followed the assignment of tasks relating to the Holzabsatzfonds. Activities are characterise with the permanent observation of home market events, monitoring of the international scene for foodstuffs playing a major role in German imports and exports, collecting and processing of significant data on different items, aggregation to and analysing of time series, distribution in bulletins with short notes, brief comments, in depth analysis and forecasts. On behalf of the Absatzfonds (German Agricultural Marketing Fund) ZMP is monitoring the market and prices of arable crops and livestock production On behalf of the Holzabsatzfonds services are provided for the forestry and timber industry. In each case this means collecting, processing, describing and evaluating of facts, figures, tendencies and trends in agricultural markets, particularly on supply, demand and prices. Presentation is done in articles, comments, tables and graphs; Dissemination in own newsletters, online services, fax mailings and as contributions to other media. Consulting and research on requests as well as. Market research concerning German and international food markets also have a home at ZMP. 127 employees work for ZMP, of which 95 are situated in Bonn, the head-quarter, and 32 in the regional and branch offices.

Index

Index

www.ingramcontent.com/pod-product-compliance
Lightning Source LLC
Chambersburg PA
CBHW081125180926
51611CB00004B/55
9789086860111